D0500048

Motivation: a biobehavioural approach

Roderick Wong
*Department of Psychology,
University of British Columbia*

PUBLISHED BY THE PRESS SYNDICATE OF THE UNIVERSITY OF CAMBRIDGE
The Pitt Building, Trumpington Street, Cambridge, United Kingdom

CAMBRIDGE UNIVERSITY PRESS
The Edinburgh Building, Cambridge CB2 2RU, UK http://www.cup.cam.ac.uk
40 West 20th Street, New York, NY 10011-4211, USA http://www.cup.org
10 Stamford Road, Oakleigh, Melbourne 3166, Australia
Ruiz de Alarcón 13, 28014 Madrid, Spain

First published 2000

Printed in the United Kingdom at the University Press, Cambridge

Typeface Swift 9/13pt *System* Poltype® [VN]

A catalogue record for this book is available from the British Library

ISBN 0 521 56175 2 hardback
ISBN 0 521 56727 0 paperback

Contents

Preface and acknowledgements

I wrote this book as a text for intermediate and advanced level courses in motivation, and as supplementary material for courses on comparative psychology and biopsychology. Although there may be overlap with material in other current textbooks on motivation, the approach and treatment taken in this one is quite different. It does not present an exhaustive review of facts and anthology of theories in the field, but instead, attempts to cover selected material linked in a coherent fashion. In doing so I have attempted to make some sense of the diverse range of topics that are covered in other motivation texts. I have also attempted to indicate the interplay of material on animal and human research, and hope that the reader will find the presentation a natural one in which the transition between the two appears unforced.

The organisation of each of the substantive chapters begins with a consideration of 'classic' theories and studies of a specific motivated activity, and is followed by discussion of selected current developments indicating further complexities of the issue. Even though some earlier theories have been superseded by recent models, like Mook (1996), I believe that students will benefit from such exposure, and consequently, develop a better understanding of how current models and research evolved. Shortly after I had completed this book, I encountered others offering new insights that were not available during my preparation. Within the limited time remaining for the production of this book, I have attempted to fine-tune some of my presentation with some ideas that I have learned from these new works. I hope that I have done justice to material gleaned from the recent books by Rolls (1999), and Schulkin (1999), as well as those from selected chapters in *Fundamental Neuroscience* edited by Zigmond, Bloom, Landis, Roberts and Squires (1999), and offer my apologies to the authors if I have been remiss. I also apologise to other excellent researchers who may feel slighted if I have not properly attended to their work. Their research is undoubtedly as important and significant as the ones that I have mentioned, but my coverage has been limited by time and space.

The reader will notice variability in the length and treatment of topics in the chapters. Part of this reflects the state of current interest and activity on the topic. For example, discussions on evolutionary and physiological/psychological aspects of sex have received much attention in recent years, and I have attempted to convey some of this material. In contrast, I was not able to devote comparable space to the analysis of social motivation, due in part to the paucity of sources that dealt with this topic from a biological perspective. However, I have adopted this perspective to make some sense in analysing the richness of the phenomenon manifested in 'real life events' *a propos* social attachment and pro-social behaviour.

An older reader may notice a superficial resemblance between the present work and one written almost a quarter of a century ago (Wong, 1976). A close examination of the contents of these two works will reveal commonalities as well as considerable differences. Although the issues remain the same, there have been significant changes in the way they are conceptualised. The influence of evolutionary psychology and advances in the neurosciences has drastically altered the direction of research on motivational issues. The experimental techniques developed in the latter field have provided researchers with means of studying proximate mechanisms, and resulted in insights that were unimaginable during the previous era.

In the preface of my earlier book, I acknowledged the help of three senior scholars in the motivation area who had read its first draft, and who brought problems to my attention, corrected some of my errors and were very encouraging. I wish that they were still around to interact with me on this project, and it is with feelings of sadness and loss, that I dedicate this work to the memory of Professors Bob Bolles, Wladek Sluckin and George Wolf. They were my mentors during my formative years and I treasured their friendship.

In its present form, this book has benefited from the support of many people. Foremost among these is my wife and best friend, Bernice, who by her own example, set standards for industry and commitment in writing. I also am indebted to my Commissioning Editor at Cambridge University Press, Dr. Tracey Sanderson, for her considerable assistance on this project. She read the first draft of the first two chapters, and spent a considerable time giving detailed comments on my errors in spelling, grammar syntax, as well as knowledge in biology. Her prompt replies to my calls for help, advice, and support throughout this project sustained me during periods when I had experienced problems on this project. During the final stages of this I received invaluable and careful help from my copy editor, Rita Owen, in the production of this work. She commented and made suggestions on material needing clarification as well as ensured that the citations and the reference list were consistent with one another. It was a pleasure to work with her, and I thank her for her forbearance throughout.

I have benefited from almost 40 years of intellectual exchange with John Jung of California State University at Long Beach, who is at present writing another of his many books. Through John's help and encouragement, I learned how to use computer-based bibliographic tools, and during this project, he walked me through many crises related to their use. Another good friend and colleague, Fred P. Valle, generously offered me the use of diagrams that he developed as visual aids for his lectures on the mechanisms and models of thirst. But most of all, I thank him for learning much from our discussions during the 25 years that we spent co-teaching a course in Animal Behaviour.

Tony Phillips was Head of the Psychology Department when I began this book, and his administration provided me with the psychological environment that made it possible. Through his postgraduate students who participated in my seminars on the biopsychology of motivation, I learned about advances in dopamine research in the Phillips laboratory and they brought to my attention, the ideas of Kent Berridge of the University of Michigan on wanting and liking as distinct motivational variables. Kent read a draft of the chapter on Feeding Activities, and offered some suggestions that I have implemented. Similarly, I thank Alison Fleming of the University of Toronto for doing the same on a draft of the maternal/parental behaviour chapter, as well as for her friendly encouragement. Richard Porter of Vanderbilt University and Laboratoire de Compartement Animal provided me with similar valuable input and psychological support on this chapter. Gary Kraemer of the University of Wisconsin helped me with information on attachment and maternal behaviour, and I thank him for his communiqué. I give special thanks to Deborah Yurgelun-Todd of the McLean Hospital Imaging Center for clarifying and amending on very short notice my description of her MRI research. Her work provides a novel and creative method of uncovering mechanisms of parent–offspring conflict. Georg Schulze of the University of British Columbia helped reorganise my arguments in the concluding chapter, and I thank him for his suggestions. Although I acknowledge the help of these people, they are not responsible for the errors. Any errors in this book are mine. I also wish to thank the following for their generosity in offering me the use of their figures or for allowing me to modify them for my own purposes. These include Edmund Rolls of Oxford University, Roger V. Short of Monash University and Edward Stricker of the University of Pittsburgh. The graphics were done by Pouneh Hanjani and Toni Ignacio, who also edited my first draft. To all of these people, my gratitude.

1

Introduction and perspective

WHY THIS BOOK?

This book is concerned with the analysis of motivated behaviour from a biological perspective. Although some psychology students may find biological topics less to their personal tastes than material that is specifically human-oriented with a social emphasis, I hope that they may be pleasantly surprised by the material in this book. It is possible to link these topics and it has been attempted in a third year undergraduate Motivation course which I taught from such a perspective at the University of British Columbia, Canada, for over 30 years. The encouraging reactions of these students during lectures and in their course ratings has motivated me to share some of this material with you.

Texts by Colgan (1989) and Toates (1986) have focused on some issues in animal motivation which form the corpus of the present book, but these earlier books were relatively short and very selective in their coverage. Although this book adopts a conceptual framework similar to that developed in the Colgan and Toates books, it is less restrictive and thus appropriate for a broader based Motivation course. Most of the recent texts on motivation (e.g. Franken, 1994; Mook, 1996; Petri, 1996) are expansive, eclectic and almost encyclopaedic in their coverage of topics. Although the framework of this book is derived from the animal motivation tradition, it can also be used to analyse relevant issues in human motivation. Thus this work sits between the larger omnibus motivation texts and the smaller ones that focus specifically on animal motivation.

Almost every text devotes a chapter, or at least a section of one to the history of motivation, and then lists various definitions of this concept by different theorists. I will not revisit this territory because it has been amply explored in many excellent sources. If you wish to visit this area, see the texts by deCatanzaro (1999, pp.1–18), Franken (1994, pp. 4–14), Geen (1995, pp. 1–19), Mook (1996, pp. 25–53) and Toates (1986, pp. 21–34). Unfortunately, definitions of motivation abound and are linked to the particular theory or model espoused by the author. 'The way that the term "motivation" is used depends upon the purpose of one's explanation, which in turn will reflect the model of behaviour that one employs' (Toates, 1986, p. 17). These differences in the scholars' theoretical perspectives not only influence their definition of motivation but also the domain of research and focus. Such considerations provoked Petri (1996, p. 22) to surmise that 'because motivation cuts across so many different subfields of psychology, its study often seems to be disjointed, with the various theories having little relationship'.

Petri (1996) dealt with the dilemma of 'disjointed electicism' by adopting an approach that focused on determining 'how differing motivating factors (biological, behavioural, cognitive) interact to produce behaviour'. This appears to be a reasonable approach and has been adopted in other recent texts in the field (deCatanzaro, 1999; Franken, 1994; Mook, 1996). However, an examination of the content of such books reveal an eclecticism in which the relationship among chapters is not always apparent. Petri (1996, p. 22) admitted that the content of his book was eclectic but suggested that the multi-factor approach provided a unifying theme. Although the recent textbook by deCatanzaro presents material from the multi-factor approach in its inclusion of evolutionary, physiological, developmental and social perspectives, its organisation and coverage are different from that of the present work.

In the present book I have attempted to avoid the problems of 'disjointed eclectism' by organising the material in a systematic manner. Each chapter will be devoted to individual systems underlying specific motivational states that result in motivated acts such as mating, parental care, feeding, drinking, exploring, affiliating, withdrawing, attacking and the like. Within each chapter as well as among the chapters I will examine the similarities as well as interactions between various systems classed as motivational. In the final chapter I will discuss whether there are any general principles of motivation that may be discerned from the analysis of these individual motivational systems. This organisation and perspective is congruent with the multi-factor approach and you will see many instances of its application throughout the book.

MOTIVATION AND LEVELS OF EXPLANATION

Toates (1980; 1986) articulated the premise that motivational factors are responsible for the organism's goal-directed commerce with biologically important incentives. This premise may provoke you to ask what is meant by the terms 'motivation', 'goal-directedness' and 'incentives'? I will answer these questions in the course of this chapter. Although experts in the field argue heatedly about definitions of these concepts and some even wonder whether such concepts are necessary (Kennedy, 1992), I believe that there is general agreement about the involvement of motivational analysis in causal explanations of action. When we ask why a person or an animal engages in a specific set of actions, we are asking about their motivation (Mook, 1996, p. 4). Motivation has to do with the reasons underlying behaviour. These reasons can be analysed on at least two levels. We can ask *why* an individual exhibits certain activities and also *how* these activities came about. The explanation of behaviour in terms of motivational mechanisms at the former level is referred to as *ultimate causation*, whilst explanation at the latter level is referred to as *proximate causation*.

Explanations of why some behaviour is exhibited implies a historical basis for the action. There may be some good reasons for its occurrence and it may have been adaptive for an animal or person to act that way. However, this adaptationist viewpoint has its critics (e.g. Gould & Lewontin, 1979). For those who favour adaptation-oriented explanations, an analysis of behaviour in terms of ultimate causation is also regarded as a *functional* explanation. What function was served by the occurrence of that specific behaviour? It is assumed that there was something to be gained by behaving in that way. Animals or humans who did not show such reactions would be unlikely to adapt as well as those who manifest this characteristic. Questions about the ultimate function of behaviour patterns in terms of adaptedness have been raised by those who are influenced by an evolutionary orientation (e.g. Alcock, 1998). Such theorists contend that the conditions of our pre-history and history influence motivational dispositions that in turn, are selected for through anatomical, physiological as well as psychological mechanisms. They assume that natural selection acted as a designing agent in shaping the proximate physiological and psychological control mechanisms responsible for behaviour. These mechanisms provide an explanation concerning *how* certain activities occur.

Many books on motivation focus mainly on proximate mechanisms at the physiological level (e.g. Mogenson, 1977; Pfaff, 1982; Stellar & Stellar, 1985) or at the psychological and social levels (e.g. McClelland, 1987; Mook, 1996; Weiner, 1980, 1985). The latter analyse the influence of motivational states or disposition on behavioural and cognitive manifestations. Most authors make passing and cursory reference to the importance of evolutionary factors on motivation and they seldom go beyond paying obligatory homage. They devote scant attention to any discussion on why motivated activities have emerged in their present form and seldom consider their implications on the selection and shaping of proximate mechanisms. Such questions and considerations have been raised by Buss (1999), Cosmides & Tooby (1995) as well as Daly & Wilson (1983). These authors present an alternative approach in which the principles of natural selection serve as a starting point for the development of models of the adaptive problems that the species of interest had to solve. Considerations of ultimate causation will be evident throughout the motivational analysis of behaviour in this book.

An example: analysis of taste preferences

I present an example that will illustrate the utility of the type of motivational analysis advocated in this book. Most of us enjoy snacks even though we may not feel hungry. What are favourite snack foods? Candy or chocolate bars? Ice cream ? Chips sprinkled with lots of salt? Salted peanuts? When we eat, foods stimulate taste receptors on the tongue and the basic taste qualities are sweet, salty, sour or bitter. It is a common observation that most animals and humans show a positive preference of foods conveying specific basic taste qualities. There seems to be a universal 'liking' of food substances that taste sweet or salty. Specific receptors on the tongue are coded to respond to sodium or sugars in foods that are ingested. When the 'sweet' or 'salty' receptor is stimulated, we exhibit behaviour that indicates a preference for foods that contain chemicals that stimulate these receptors. We show continued ingestion of such foods and experience a positive psychological affective state or hedonic experience. In fact, the Greek word for pleasure, *hedone*, relates to the Greek for honey, *hedus*. Booth (1994) proposed that there is a pleasing character of the activity of satisfying a desire for a sweet food. The same may be said about our liking of slightly salty foods. What are the reasons for this reaction?

The taste affect is 'hard-wired' since it is present before we, as infants, have had any experience of the consequences of ingesting sweet substances such as sugar. This inborn affect of taste explains why we use terms involving taste to express value. We often comment that a person has a 'sweet disposition' or that another person has a 'bitter outlook on life' (Bartoshuk, 1989). These are examples where we have borrowed the affective tone of the taste quality and applied it to some other affective situation. Biological factors may be invoked to provide a reasonable explanation. Taste is the sense that is tuned in to nutrients. A salty taste is characteristic of sodium whereas a sweet taste is characteristic of biologically useful sugars. In contrast, the bitter taste is characteristic of a variety of poisons, whilst the sour taste is associated with highly acid and sometimes rotting foods.

A motivational analysis of the reasons for why we seek and ingest quantitities of sweets and salty snacks considers both *ultimate* as well as *proximate* causal factors. Evolutionary factors are responsible for the selection of physiological and psychological mechanisms that are responsible for our reactions to sweet and salty tastes. This consideration would spur us to inquire into the nature of human salt and sweet preference mechanisms and whether they mesh with our physiological requirements for salts and sugar and what the opportunities were for procuring them during our pre-history (Cosmides & Tooby, 1995). I will elaborate on this analysis in greater detail in Chapter 5.

MOTIVATIONAL PROCESSES AND PROBLEMS OF REDUCTIONISM

Earlier, I had referred to Toates's (1980, 1986) notion that motivation involves goal-oriented commerce with incentives that are usually of biological significance. When an organism is motivated to behave in a certain way, this behaviour is terminated when a goal is achieved. In that respect, motivated behaviour may be characterised as guided by its consequences and is related to some end point linked with biological requirements of the organism. McFarland (1989) argued that although motivated acts appear to be directed by goals, these acts are not necessarily specifically directed toward some internal representation of that goal. The motivational state persists until the goal has been attained. The success of this motivated act results from either a change in the organism's relation to the external world, a change in its internal state or both. I will return to this discussion on purposive behaviour and goal-direction in the concluding chapter.

Motivational processes are inferred from properties of behaviour that appear to terminate a sequence of goal-oriented events but cannot be observed directly. Thus, motivation is a dispositional variable or concept that is inferred from behaviour. Whenever we analyse motivation or motivational processes, we accomplish this by studying motivated behaviour. Although physiological interventions may influence the state of the organism, the main focus of our observations is still the behaving organism. For example, no one has ever seen *thirst*. We can observe how long a person has gone without water and then assess how much he drinks, or how hard she will work or pay to get some water. But we can never see thirst itself. Even if we probe inside the person's head, we could see only brain cells, not motives. The physiologist adopting a reductionistic approach seeks an explanation of behaviour in terms of neural activity and must have some idea of about the properties of the behaviour he or she is trying to explain.

In 1951, Tinbergen, a pioneer in ethology, a field dealing with the biology of behaviour, pointed out the problem in bridging the great gap between what nerve cells do and how animals behave. Tinbergen (1954, p. 115) stated that 'to try to arrive at an understanding of the causation of behaviour by jumping to the level of the neurone, or of simple neurone systems is extremely harmful . . .'. It is harmful because 'the machinery is so complex that looking at the millions of parts that go to make it up will only confuse and make it more difficult to see how the the whole works' (M.S. Dawkins, 1986, p. 98). Even more critical is the fact that 'the complete wiring diagram of the nervous system would not constitute understanding of how behaviour works. Real understanding will only come from distillation of general principles at a higher level' (R. Dawkins, 1976, p. 7). Following Tinbergen, these ethologists adopted the view that the wrong way to determine behaviour mechanisms is to try to work from the bottom up. Behaviour must be approached at its own level rather than simply as a projection of physiology (Halliday & Slater, 1983, p. 4). Thus the physiologist must be guided by considerations of the manner in which thirst is manifested.

ORGANISATION OF THIS BOOK

As I mentioned earlier, the chapters in this book are organised around specific forms of motivated behaviour. In objective terms, such behaviour involves molar acts that regularly consummate or terminate a recurring behaviour sequence related to some dispositional state. Acts such as mating, caregiving, feeding, drinking, sleeping, 'dreaming', exploring and withdrawing from and attacking aversive stimuli are consummatory in structure. The functional significance of these acts is quite apparent. The appropriate execution of these acts maintains the individual's fitness and involves vital biological, psychological and social processes. The concept of *fitness* will be discussed in detail when I deal with the analysis of mating or reproductive behaviour in the next chapter.

Each of these motivated acts will be analysed in the subsequent chapters from the perspective of ultimate and proximate causal factors.

Ultimate causal factors involve the process of natural selection that is responsible for the retention or elimination of specific characteristics over successive generations. This process results in the selection of genes guiding the developmental programmes constructing mechanisms to produce behaviour. Thus behaviour is an effect that is an outcome of a causal system that reflects the effects of proximate mechanisms. As a result of the effects of natural selection that shape these proximate mechanisms, the resultant behaviour correlates to some extent with the organism's biological fitness. In general, the evolutionary function of the brain is to process information in ways that lead to adaptive behaviour.

The success of an organism depends upon the existence of a set of interacting motivational causal systems that enables it to deal with the major problems of survival and gene propagation. Colgan (1989) grouped these causal systems into various domains or areas. The primary domain concerns *reproduction* and entails successful courtship and parental care. The second grouping concerns *nutrition and fluid balance* whilst the third deals with *personal care* which involves sleeping and care of the body surface. The fourth category involves *agonistic reactions* and corresponding motivational states of fear and aggression are manifested in behaviours such as fighting and fleeing. Colgan's grouping of causal motivational systems is logical and makes much sense. I have adopted the basic structure of Colgan's taxonomy of motivational causal systems but have omitted discussion of the personal care system. Instead, I have extended the behavioural categories in some domains as well as added an additional category or major grouping dealing with social motives.

A preview of the forthcoming chapters

I have arranged the sequence of material in the subsequent chapters in terms of specific motivated activities. They belong to one of the causal systems in Colgan's (1989) list of basic functions as well as one concerning social motivation. The most essential property of life is reproduction and from this premise it follows that the primary motivational causal system is concerned with this domain. Chapter 2 outlines various aspects of sexual behaviour that range from evolutionary to physiological to psychological to social and cultural factors. In this chapter I examine how biological characteristics of the sexual reproductive system are manifested in male–female behavioural differences in animals and humans. In addition, I consider current thoughts concerning the fitness interests of males and females and the reasons for their mating tactics. In this context, I examine recent material on cross-cultural studies on human mate choice. I also analyse recent research on the ontogeny of male–female differences from both the physiological and social perspective.

In most sexually reproducing species, it is not enough that males and females engage in successful mating encounters that result in the production of viable offspring. The fitness interests of these animals are enhanced by parental caregiving to the offspring, usually by the mother. Chapter 3 deals with a topic that has not received the coverage and treatment which it deserves in other texts on motivation. In this chapter I consider the importance of this behaviour from an evolutionary perspective and examine examples from mammalian and avian species. I also analyse research on human parental care from this perspective. In addition to the analysis of parental behaviour at the level of ultimate causal mechanisms, the influence of proximate causal mechanisms at the physiological and psychological levels will also be examined.

Chapters 4 through to 6 deal with aspects of the grouping of motivational causal systems dealing with nutritional and fluid balance. In Chapter 4 I examine the concept of homeostasis and relate it to properties of the feeding system. Most reviews on feeding tend to regard this behaviour as a mere endpoint for the evaluation of the effectiveness of various physiological manipulations. Although material from this position is covered, I consider a synthesis of the physiological, psychological and ecological aspects of the feeding system. In this synthesis I examine the interplay of energy/repletion cycles and homeostatic mechanisms as they relate to the role of environmental/psychological factors in the regulation of feeding. I also examine two types of eating dysfunctions prevalent in some humans, that of obesity and anorexia. In this analysis I deal with social, psychological as well as physiological variables together with their interplay but in addition, I consider possible evolutionary factors that may influence the dispositional tendencies of some individuals to fall prey to one or other of these dysfunctions.

Chapter 5 deals with the other important aspect of feeding, the selection or choice of nutrients. Most of the existing books on motivation have focused mainly on factors that initiate, maintain and terminate feeding, concentrating on consumption and rarely dealing with food choice or selection from an array of possibilities. The issue of food selection and its causes is an important one because the integrity of the system is dependent upon the appropriate selection of mechanisms. I examine how factors operating at the physiological, psychological and social/cultural levels influence our choice of foods. The ultimate causal or evolutionary mechanisms responsible for the manifestation of these specific proximate mechanisms is brought into the analysis. This is of particular interest when we study the strong influence of culture and cuisine on food preferences in humans.

Chapter 6 concerns fluid regulation within the body arising from the interaction of internal and external factors. Drinking may be viewed as a behavioural mechanism for either maintaining or restoring the water balance of the body. Solutes within the body tissues and deviations from an ideal solutes ratio are detected by specific receptors in the nervous system which in turn activate behaviour that directs the organism to seek fluids. Physiological mechanisms also regulate the internal water balance by acting upon physiological mechanisms that prevent further loss of water through internal conservation. In addition to these factors, fluid intake can be instigated when the solute/solution ratio remains unchanged but there is a loss of absolute volume in body fluids, such as may occur following severe blood loss, or loss of liquid arising from diarrhoea or vomiting. Drinking, however, is also influenced by factors other than that of the internal state of the animal. Palatability of the liquid and its effects on the hedonic state of the animal is an important factor affecting both intake and choice. Cues from the external environment also affect the motivational state of the animal and results in enhanced intake among animals whose water balance is normal.

All animals have the tendency during a portion of their waking moments to explore their environment and engage in stimulus-seeking activities. Chapter 7 will deal with this type of motivated activity, one that is provoked by the interaction of the state of the nervous system and the stimuli that are impinging upon its sense organs. Behaviour of this sort involves responses toward novelty and is believed to represent a possible information-gathering mechanism essential to the animal's adaptive adjustment to its spatial environment. Unlike the other motivated activities, exploratory behaviour is believed by some motivation theorists to be independent of drive-reduction and reinforcement. Because of this consideration, Green (1987, p. 318) has argued that such behaviour 'can be seen to lie completely outside the traditional approach to the motivation of behaviour'. In this chapter I will present data and theory that suggests the differences in mechanisms may not be as great as Green suggested. In addition, I will discuss the relationship between exploration and fear as as well as Zuckerman's (1994) integration of biological and physiological considerations to the analysis of sensation seeking tendencies in humans.

When animals are exposed to aversive stimuli, they are likely to display either escape or attack behaviour. Both types of responses to the source of aversiveness have adaptive significance. Specific escape mechanisms have evolved for dealing with physical danger. One of the simplest is the withdrawal reflex that removes the organism from damaging stimuli. For example, taste receptors that respond to bitter or very sour substances elicit a spitting reflex as protection against those chemicals which ingested would have dangerous consequences. More complex situations that elicit such protective and defensive reactions will be elaborated in Chapter 8. The analysis implicates psychological mechanisms that enable animals to anticipate threatening and damaging events and thereby avoid contact with them. Animals will also react to aversive stimuli with attack behaviour. In many respects withdrawing and attacking behaviour can be analysed in a manner similar to that applied to the other consummatory activities. In this chapter I indicate the value in analysing the behavioural effects of aversive stimuli in terms of responses elicited by them. This contrasts with the learning framework that focuses more on the fact that animals acquire responses that are instrumental in removing themselves from the source of aversion. Such responses become strengthened and integrated as a result of repeated exposures in the test situation and are regarded as acts developed through learning. In contrast to this interpretation, animals may be pre-programmed through natural selection to acquire specific responses because they are elicited by aversive stimuli. The instrumental aspect of the situation may be fortuitous. From the perspective of proximate analysis, the instrumental aspect of the situation focuses on the function of the behaviour whilst the elicited aspect focuses on the causation of the response.

Of course there is much more to the analysis of responses to aversive events than elicited reactions. Motivational and emotional states are induced in such situations and they are presumed to mediate the direction which behaviour takes. Psychological states such as fear or aggression are the mechanisms that determine whether an animal expresses either 'flight' or 'fight' in its reactions to stimuli in its environment. The context in which the animal encounters these stimuli as well as its previous history with them influences the 'interpretation' that it applies to the situation. Physiological and psychological experiments on these states and the circumstances of their instatement will be presented in Chapter 8. The chapter will also analyse a source of aversion that does not involve pain, namely that of frustrative non-reward. In this analysis we will encounter Gray's (1987) model that demonstrates the functional similarity of mechanisms underlying fear and frustration. A discussion of aggression from both an evolutionary as well as the physiological/psychological perspective will end this chapter along with the issue of homicide and recent evolutionary psychological explanations.

Some behaviours that reflect the manifestation of social motives such as attachment and altruism will be discussed in Chapter 9. There is considerable literature analysing these motives from the perspective of social and developmental psychology as reflected in the recent texts on human motivation mentioned in the beginning of the present chapter. In this short chapter I will revisit this territory but focus on the biological foundations of these behaviours. During the 1990s there has been interest in applying both proximate and ultimate causal analysis of these phenomena. Although much of the research has been concerned with animals, representative studies on humans arising from theories coming from animal studies are more prevalent. This chapter will indicate that there is a consummatory element to these motivated activities just as there is to those that are more directly concerned with the integrity of the individual system. However, in concert with the theories derived from the social and developmental psychology framework, I discuss the extent to which processes and mechanisms of attachment and altruism among humans may reflect variables of a different order than those of the other species.

The final chapter of this book is concerned with the commonalities, differences, as well as interactions of the motivated activities that have been studied in the previous chapters. Motivated behaviour patterns are the products of an intimate interaction between the organism and its environment. In the analyses of individual motivational systems covered in each chapter I have attempted to relate physiological with psychological events not by reducing or explaining behaviour in physiological concepts but by indicating the reciprocal interaction of physiological processes and response patterns. Thus biological events are an essential part of psychological events but are no more basic to behaviour than are other factors such as the characteristics of the stimulus object or the organism's past interactions with a stimulus object.

In Chapter 10, I will indicate how some of the material of the preceding chapters are related. For example, it is obvious that feeding and drinking are interrelated. When an animal ingests food, it also takes in water that appears to anticipate later food intake. Meals influence the timing and amount of drinking because food introduces solute into the blood and shifts water from body fluids into the gut. However, feeding can be inhibited by variables that affect the other consummatory acts. If an animal suffers a water deficit as a result of water deprivation, it will voluntarily limits its food intake. Of course, when the animal is allowed access to water, its food intake will be increased. Also feeding is momentarily inhibited when an animal is fed in a novel environment and during a different period of time. The animal requires a period of adjustment to the new feeding conditions before it is able to exhibit compensatory feeding behaviour. The novel stimuli in the test situations elicit stimulus seeking or exploratory behaviours that interfere with feeding. The animal eats once it habituates to the stimuli in the test apparatus and the stimuli are no longer novel to it. In other words, the animal no longer displays neophobic reactions. A similar treatment will be applied to the interaction of other motivational processes.

In the concluding part of Chapter 10, I consider recent developments which resulted in a re-conceptualisation of rewards and their effects on motivated behaviour. The recent work of Berridge (1996) has provoked a re-examination of traditional views concerning reward and its relation to brain systems. His research suggests that reward contains distinguishable psychological components consisting of an affective component (liking) and an appetitive or incentive component (wanting). These can be manipulated and measured separately, and Berridge's research also indicated liking and wanting processes arise from vastly distributed neural systems. His work suggests a way in which affective or hedonic factors (*pleasure*) and motivation may be instantiated by the interplay among neural systems. The manifestation of motivated behaviour may thus depend upon the separate but equally important contributions of *affective* and *incentive* processes.

SUMMARY

Motivation is a dispositional variable or concept that is inferred from behaviour. Motivation concerns the reasons underlying behaviour, and can be analysed on at least two levels. We can ask *why* an individual exhibits certain activities (ultimate causation) and also ask *how* these activities came about (proximate causation). When an organism is motivated to behave in a certain way, behaviour is terminated when a goal is achieved. Such behaviour is guided by its consequences and is related to some end point linked with the biological requirements of the organism.

The first nine chapters in this book are organised around specific forms of motivated behaviour, which are analysed from the perspective of ultimate and proximate causal factors. The material in these chapters also indicate how the manifestation of motivated behaviour is dependent upon the separate but equally important contributions of *affective* and *incentive* processes. The final chapter deals with the relationship of material of the preceding chapters, and suggests how affective factors and motivation may be instantiated by the interplay among neural systems.

2

Mating and reproductive activities

Until recently most psychologists studying reproductive behaviour focused on proximate causal mechanisms and paid scant attention to issues concerning the functional aspects of such behaviour. There are some outstanding exceptions. Although Donald Hebb is best remembered for his seminal work on physiological models of learning, memory and other cognitive processes, his views on evolution and behaviour also revealed foresight in the following statement. 'The function of behavior in evolution is simply to keep an animal alive and well enough to mate and in other ways to get the next generation established, the process then repeating itself and leading to still another generation' (Hebb, 1972, p. 171). Few psychologists during that era considered ultimate causal mechanisms in their theoretical analysis aside from short passing reference to the significance of evolutionary factors in the explanation of behaviour. They seldom ventured beyond that level. Textbooks on the psychology of motivation with an experimental emphasis tended to focus on proximate causal mechanisms underlying rat behaviour (Beck, 1978; 1990; Bindra, 1959; Bolles, 1967; Brown, 1961; Hall, 1961; Young, 1961).

Following the publication of E. O. Wilson's *Sociobiology: the new synthesis* (1975) and R. Dawkins' *The selfish gene* (1976), a new group of psychologists were inspired by the implications of evolutionary factors to the explanation of motivational processes. This 'new generation' of behavioural scientists which include Buss (1989, 1999), Cosmides & Tooby (1987; 1992; 1995), Daly & Wilson (1983; 1988; 1994) and Symons (1979) have reconceptualised motivational analysis through their incorporation of evolutionary theory. This approach is now referred to as evolutionary psychology and reflects the diverse intellectual background of this new generation of researchers who were originally trained in biology, ethology, anthropology, as well as comparative and physiological psychology. Along with contributions from other life scientists of their generation, this group has contributed considerably to our current understanding of sex as the basic motivational force in behaviour. In the following sections I summarise the major ideas of this group of researchers.

Amongst motivated behaviours, reproductive activities are directly linked with self-perpetuation, whilst other motivated activities are subsidiary to this outcome. Evolutionary psychologists have proposed that living organisms are selected with proximate mechanisms that were linked with reproductive success in their ancestral form (e.g. Tooby & Cosmides, 1990a, b). However, reproductive success is *not* a cause or a goal of organisms, but an *effect* of action. It is an effect of the ancestral form's actions and *not* that of the present living organism. The motivation to pair successsfully and as often as possible does not necessarily reflect an intentionally reasoned approach but instead is the consequence of a mindless evolutionary rachet – success breeds success. The behaviour is merely a consequence of proximate mechanisms, such as the urge to mate, that has been selected for, and conscious intent need not accompany this sequence of events.

THE IMPORTANCE OF SEX

Genes are passed from one generation to the next and their passage is affected by the behaviour of the animals carrying them. Motivated activities such as ingestive behaviours, dealing with predators or building shelters are important because they enable animals to achieve reproductive status and thereby perpetuate some of their genes. However, the primary method by which organisms ensure genetic prosperity is through the motivational force regulating reproductive activities. Richard Dawkins (1976) regarded the body as a loose confederation of 75 000 pairs of genes, all of them dependent upon each other for survival but each competing for representation in the next generation. What are the means by which the organism can perpetuate its genetic material? There are two possible methods of reproduction, sexual and asexual. With the asexual method, the organism simply produces exact genetic replicas of itself. A cell divides, a section buds off, a clone results, and one has complete genetic replication. In contrast, sexual reproduction is a more complicated and wasteful method.

Sexual reproduction involves a process where two individuals of the same species combine their genetic material to produce a novel genotype. Each parent contributes half of the genes of an offspring. Two steps are involved in this process. Gametes or germ cells are produced by meiosis, a process which produces a single haploid copy of each chromosomal pair in the male and female gametes. After the two gametes (a haploid egg from the mother and a haploid sperm from the father) fuse, a new diploid zygote is formed. The new organism usually has two versions of every gene which are organised into long strings called chromosomes and they exist in pairs, except for the sex chromosomes. This form of reproduction creates more variability in each generation and so improves the chance of a greater number of individuals in a species surviving in an ever-changing environment.

It may seem that asexual reproduction, involving the cloning of offspring from a parent, is more efficient than the sexual method in ensuring that all the genes are passed on (Fig. 2.1). This may make one wonder why so many animals reproduce sexually rather than asexually. After all, sex is expensive. Besides expending energy in the development and use of the necessary organs, sex permits an individual to pass on only half of its genetic material to its offspring. If natural selection maximises gene transmission, why should sexual reproduction ever evolve? Theoretically, evolution should favour asexual organisms that pass on their entire genetic complement. What advantage is there in sexual reproduction? This question was a puzzle to Charles Darwin (1862) who proclaimed that 'the whole subject is as yet written in darkness'. This query prompted subsequent scholars to offer their solution of this question. In the normal course of things, simple microorganisms reproduce themselves quickly and accurately when conditions are right. However, when they encounter difficulties in finding food, or suffer from predation or poisoning, they often abandon the asexual method of reproducing and engage in sexual reproduction.

Early explanations suggested that it was 'for the good of the species' to have the variations that are brought on by the mixing and shuffling of genes inherent in sexual reproduction. This method enables a population to have a better chance of having some individuals who would have the right genes to be able to 'cope' with some future environmental change even though others might survive through the less costly asexual method. Although this seems to be a plausible reason, it has a major weakness. A trait as costly as sexual reproduction would not persist if it were detrimental to its bearers. Other theorists suggested that when an organism lives in changing environments, including that made up of other organisms, there are benefits if it is able to produce offspring that are different from it. Competitors, prey, parasites and predators of these organisms may evolve their own adaptations and present changing selection pressures that may make sexual reproduc-

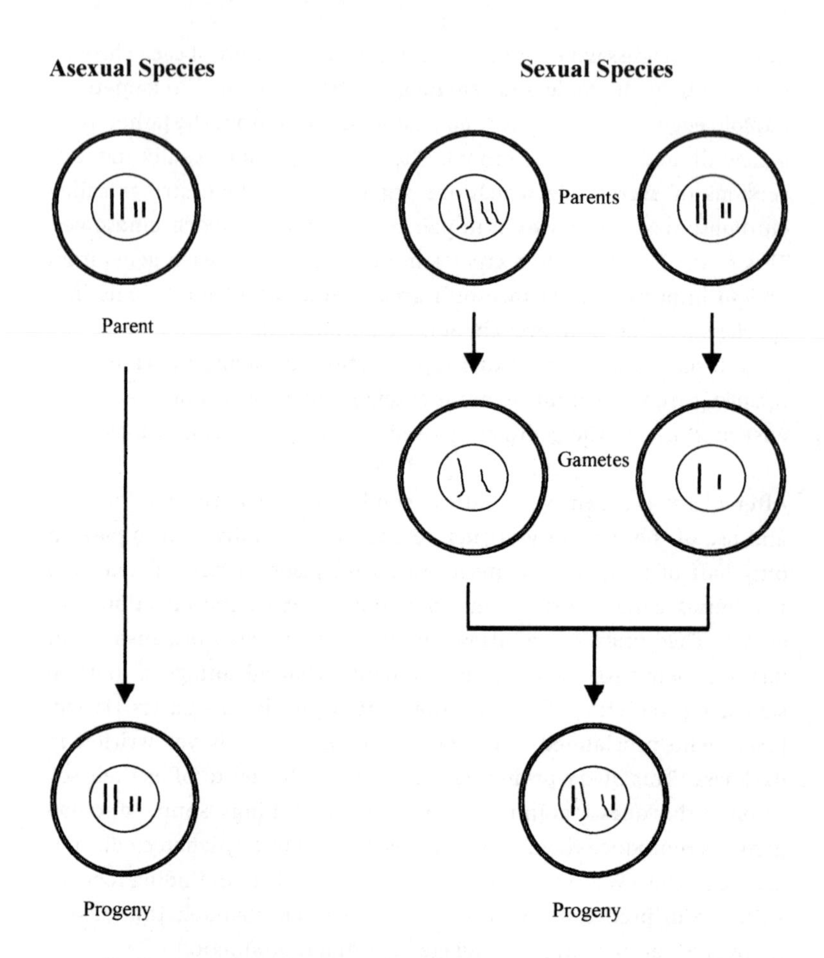

Fig. 2.1 Asexual species produce identical copies of themselves, whilst individuals in sexual species contribute only half of a progeny's genome. (Adapted from *Sexual Selection* by Gould & Gould, 1997. Copyright 1989 by Scientific American Library. Used with permission from W. H. Freeman and Company.)

tion more viable than the asexual mode (Hamilton, Axelrod & Tanese, 1990). This advantage arises from the diversity possible with sexual but not asexual reproduction. A more recent theory suggests that sexual reproduction maintains the overall fitness of a species by clearing out harmful mutations such that the least healthy offspring die, taking the damaged DNA of the genes with them. Zeyl & Bell (1997) found that sexually reproducing yeast clear out harmful mutations more efficiently than yeast that do not reproduce sexually.

SEXUAL STRATEGIES

Amongst sexually reproducing species there are two morphs or forms – male and female. Although there are some hermaphroditic species such as snails, in which individuals possess both female and male organs, they are the exception. There are profound differences in the make-up of males and females arising from morphology, physiology and psychology. Although both sexes behave in ways that can ultimately be related to matters of fitness or reproductive success, they differ in their strategies for achieving this end. Such behavioural differences can be linked to the consequences of the nature of sexual reproduction. Through this system male and female morphs within the same species combine their genetic material and a novel genotype is produced as a result.

The gametes of males and females differ greatly in structure and function. The male produces a host of microscopic and mobile germ cells called the spermatozoa in the testes in contrast to the large and relatively immobile eggs that the female produces in the ovaries. The size ratio highlights the immense difference between male and female gametes. In humans, the ovum or egg is 85 000 times larger than the sperm cell. A single egg is released approximately every 28 days in a woman whereas a man expels around 100 million sperm in each ejaculation. Intersperm competition is believed to be the basis of such astounding male–female gametic size difference. Because so many sperm are released at the same time and all pursue the same goal of searching, finding and fertilising the ovum, these cells are stripped down for mobility.

The sperm's cellular structure consist of three parts: a small conical head that contains the nucleus and its chromosomes – the blueprint for the next generation; a ring-like midsection containing mitochondria – the cell's powerpack; and a whip-like tail that propels the body. These parts are illustrated in Fig. 2.2. Once the ejaculate reaches the entrance of the female reproductive tract, the sperm swim up the fallopian tubes racing to reach the ovum that will die if it ripens before the sperm get to it. With the aid of a high-powered microscope we may observe variability in the forms of sperm within an ejaculate of human sperm. One form is a stripped down model in which the cellular material is minimised in the service of motility. It has a long tail that enhances motility but is comparatively short-lived compared to the other form. This has a larger mid-section and a shorter tail. In contrast to the first form, it is less motile but has a longer-life. Although these forms or morphs of sperm reflect designs to attain the same end point – the fertilisation of the egg –

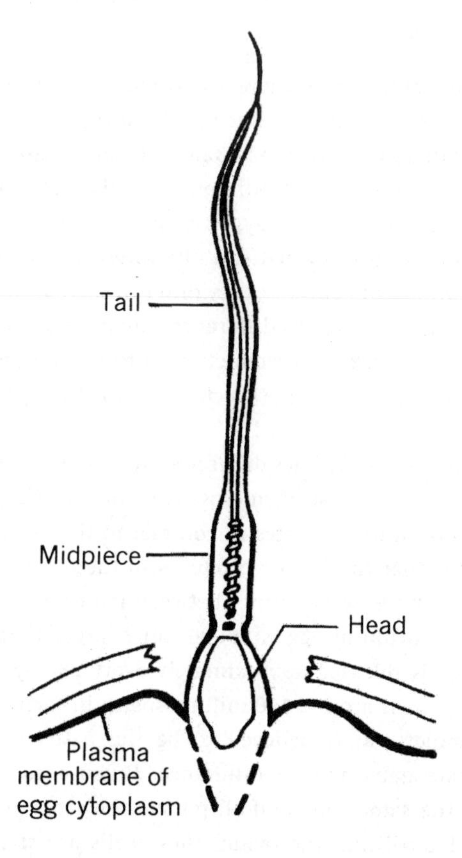

Fig. 2.2 Sperm cell consisting of tail, midsection, and head during entrance into the egg through its plasma membrane. (Drawn by Pouneh Hanjani.)

their success is dependent upon specific conditions. One concerns the timing of intercourse in relation to ovulation. If the female's eggs have been released during the period of copulation, because the smaller sperm swim up the fallopian tube (the female reproductive tract) faster, they reach the egg sooner. For this reason, they outcompete the larger slower ones. However, if ovulation occurs after copulation has taken place, the larger and less motile sperm are more likely to fertilise the egg because of the larger nutrient stores which allow them to outlive the smaller ones. Such size differences may be relevant to the analysis of sex differences.

There are indications that sperm carrying the X- and Y-chromosomes differ in morphology. Sumner & Robinson (1976) found that the mass of Y-bearing sperm was slightly lighter than ones carrying the X chromosome. Other researchers reported that the swimming speed of the Y-bearing sperm is higher than the ones bearing the X (Ericsson, Langevin & Nishino, 1973; Sarkar, 1984). The faster swimming speed of sperm carrying the Y-chromosome explains the fact that between 120 and 150 boys are conceived for every 100 girls (Purtillo & Sullivan, 1979). However, more male foetuses are lost in spontaneous abortion than females.

In contrast to the sperm, the unfertilised egg or ovum is a large specialised cell, the surface of which is studded with short and stubby projections called microvilli. Large reserves of food are stored in this cell. The number of eggs released by the female is relatively small and in humans, they are produced during foetal development and remain dormant until puberty. After puberty one egg is released on average each month from the ovary until the woman reaches the menopausal period, which occurs when the supply of eggs dries up.

The common distinction between male and female gamete activity implies that sperm are more active and vigorous in their pursuit of their goal than the larger and less active ovum. The egg's immobility is due to the bulky cytoplasmic mass that house its chromosomes. Such a distinction has been postulated to constitute the developmental basis of activity–passivity differences pervasive in adult male and female organisms. However, Batten's (1992) critique of the 'myth of the passive female' tempers this distinction. The issue is not quite that simple. When the nuclei of the male and female gametes unite during the fertilisation of the egg, both sets are fused, and the programme for the new individual takes shape. The sperm acts like a microsyringe and injects calcium ions and alkaline cytoplasm, which triggers local changes in the egg's membrane around the spot on which the sperm has landed.

When the sperm and egg first touch, a long filament shoots out of the sperm and harpoons the egg. Some of the tiny projections (microvilli) from the egg's surface extend upward and embrace the sperm. These projections clasp the sperm's head and eventually entwine its tail. To prevent other sperm from entering, the egg produces two kinds of block. The first, arising from changes in polarity of the membrane following contact with the sperm, is electrical, rapid and temporary; the second is structural and long-lasting. The electrical block lasts for about 30 seconds and is the result of the change in polarity from negative to positive. The structural block develops as a consequence of secretion of protein substances under the egg membrane that cause this region to be inpenetrable by other sperm.

Imbalance in male and female investments

When a female is ready to mate, there is usually more than one male available, all of whom are motivated to interact with her. This situation generates much competition among males and incurs costs in each of them. Males expend a lot of time and energy in competing for the prospects of mating with a female. Such dynamics are evident even at the level of gametes. Prior to gametic contact, sperm cells seem to expend more energy than the egg. However, once the female and male gametes fuse, the situation changes and the female makes a greater contribution. Her egg contains food stores and energy is expended in the cellular division of the new zygote occurring within her structure. Upon the basis of these considerations, it follows that the loss of the zygote is more costly for the female than the male.

The female has invested considerable resources in producing the egg and the time required for her to produce and release the next egg is much greater than that for the male to produce more sperm. This indicates that the female exhibits greater parental investment even before her egg is fertilised by the male's sperm. In mammals where internal fertilisation is the rule, the pregnant female nourishes the embryo during its development. Following birth the mother continues to provide caregiving and feeding of her offspring. In contrast to the female's strong commitment to the care of her young, males tend to provide minimal support. Amongst most mammals, males appear to contribute only genes to the zygote. In some respects sperm might be regarded as parasites that exploit the contribution of the egg to the development of the zygote.

We might assume that males exploit females by producing large numbers of offspring at very little cost to themselves. This imbalance in relationship, which begins at the level of eggs and sperm, is regarded as the precursor of the imbalance that exists at the adult level among mature male and female members of a species. Trivers (1972) had suggested that the bearer of the scarce egg is a scarce resource for which bearers of sperm compete. Because of this imbalance, it would follow that strategies which are effective for one sex may not necessarily work for the other. Thus different strategies to maximise reproductive success have been evolved among males and females. As a consequence of these differences, we observe the phenomena of competitive males and choosy females.

Winning the female through contests and looking good

In species where the male does not help with the care of young we might assume that it is more advantageous to be a male than a female. As mentioned in the previous section, the male faces many costs other than the minimal one of producing the single sperm that fertilises the egg. In order to achieve this, the male has to make millions of other sperm which are simply wasted. Then in order to be successful in mating, the male may have to fight to gain access to the female or to defend a territory to which the female is attracted. A good part of the male's time is spent competing with one another for such territories and resources valued by females. Among non-territorial species, males compete for access to females during breeding seasons. Fighting is a costly enterprise that even the most successful males find rewarding only during the brief period of time in which they attain this high status. Such males often encounter challenges from younger males who have matured to the level where they match the competitive abilities of the challenged.

M. Dawkins (1995) has argued that males cannot be said to be more successful than females or be regarded as exploitative if this implies that females are at a disadvantage. Dawkins pointed out that females have, on average, just as many offspring as males. Although an individual male may have a more successful reproductive history than any given female, on average, the hard-working females are just as successful as many aggressive and competitive males. Although males have a much higher reproductive potential than females, they also have a higher chance of having no matings whatsoever because of the competition among males. In contrast, females are more likely to survive and rear offspring.

Aside from pressures favouring the ability to compete directly with one another by winning fights (intrasexual selection), there is also selection pressure favouring traits in males which attract females (intersexual selection). Characteristics such as bright colours, long tails, elaborate plumes have less to do with combat and are more likely to be directed at attracting females. In his analysis of the highly ornamented peacock's tail, Darwin (1871) concluded that this character arose because of the female peahen's preference in mating with the best-ornamented males. He suggested that mate choice acts as a selective force that brings about differential rates of reproduction favouring those individuals that bear the preferred characteristics. Structures such as large and heavy antlers or horns that form the basis of female choice in ungulates also serve as natural weapons which are directed at other males. Current research suggests that such characteristics may also reflect the health of the male (Kodric-Brown & Brown, 1984) and thus it may be difficult to make a clear distinction between intra- and intersexual selection. Male competition and female choice are often inextricably bound together. It is to her benefit that she chooses a mate that is strong and healthy as well as one that can withstand aggression from other males. These characteristics are mediated through the effects of testosterone as indicated in Fig. 2.3.

Male elephant seals are larger than the females by a factor of three to eight and the males have elongated canine teeth that they employ like daggers. Males compete for special areas along the beach and broadcast loud raucous threats when challenged by other males. If their rival does not back off, the holder of a desired site rears up and and then falls on his opponent as hard as possible while driving his teeth into the other's neck. The strongest male in the social system claims the best site along the beach. The females search for what they consider the best calving site and the male that controls it may not necessarily be of their own choosing. However, female elephant seals sometimes exert an influence by assessing the fighting ability of this male. When mounted by a male, the female emits loud calls that attracts the attention of nearby males. This 'incitement' provokes bigger and stronger rivals to oust her current mate and the 'winner' then proceeds to copulate with her (Le Boeuf, 1972). Although the female does not choose her mate in the conventional sense, her behaviour does exert a major influence on which male sires her offspring.

Although there are strong arguments for regarding inter- and intrasexual selection as two separate processes, there are many examples that indicate the complexity of this issue. We have already mentioned that large horns and antlers among male deer and moose may serve as signals of good health and vigour to females. Among birds, male ornaments such as plumes or patches of brightly coloured feathers are also linked to both male health and vigour through their role in male–male interaction, as well as to female attraction. The female shows preference for mating with such males compared to their less endowed competitors and perhaps this is a response to indications that such a male has proven his worthiness in agonistic encounters with other males. Maynard Smith & Harper (1988) suggested that many bird species signal dominance through 'badges of status' such as patches of feathers or black throat patches (Møller, 1987). These are the means by which to check whether such a badge honestly indicates the quality of the male. Møller observed that birds sporting high status badges receive more

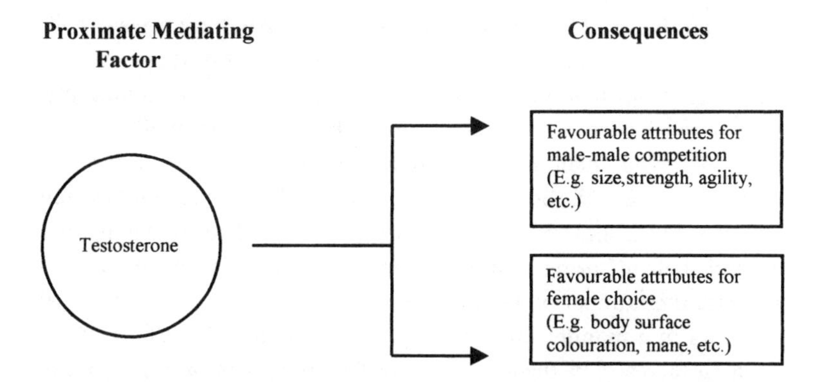

Fig. 2.3 The consequences of testosterone on attributes that facilitate success in inter-male competition and female choice.

aggressive challenges from other dominant birds and thereby pay a cost not paid by birds in subordinate plumage. Thus females that are invariably attracted to the status badge also gain a fit and healthy male. This example again illustrates the point that an event that appears to be a pure case of female choice may also be interpreted as choice being linked to male quality through male–male interaction.

PARENTAL INVESTMENT, SEXUAL DIMORPHISM AND MATING SYSTEM

We have examined how intrasexual selection exerts its influences on mammalian males by selecting for characters such as body size, mass, strength and other characteristics that enhance competitive advantage. We also are aware of the imbalance in parental investment among male and female mammalians and the reasons for it. As a consequence of these factors most male mammals can be classified as polygynous, a system in which each male attempts to mate with several females. This was documented in our discussion of the elephant seal in which the 'winners' of valuable beach sites attract many females who then mate with them. Such individuals would then have a larger number of offspring than the 'losers' and pass on genes for the strategies associated with successful polygyny to the next generation. Multiple mating attempts seem to be the norm among mammalian males.

The situation among avian (bird) males is different. Although monogamy is rare in mammals, this mating system is common in birds. Lack (1968) documented that in about 90 per cent of all avian species, each male forms a bond with a single female during the breeding season.

Although the results of studies (see Birkhead, 1996, for a review) have challenged the notion of universal fidelity among birds, they nevertheless differ from mammals with respect to prevalence of promiscuity. Why is there such a difference between birds and mammals? Unlike male mammals, male birds are more able to participate in parental activities along with the females. Only the female bird can produce and lay eggs, but males can incubate them, bring food to the young and protect their offspring from predators. A monogamous male bird can increase the number of descendents he produces through his parental activities in a manner unlikely among male mammals (Orians, 1969). Such an increase, in theory, may outweigh any potential reproductive gain he might have made were he to adopt a nonparental strategy through the pursuit of many females. A male that divides his effort between two or more females might face the situation where he produces no offspring (Wittenberger, 1981). He stands to lose more if he were to desert his mate after copulating with her. Female birds benefit greatly from the aid of their mate in the rearing of their offspring. Those without male aid rear fewer young than those that have more aid (Lyon, Montgomerie, & Hamilton, 1987). In contrast, female mammals are usually successful in rearing young even when males desert them to exploit other mating opportunities.

Morphological and behavioural differences between the sexes are referred to as sexual dimorphisms. Darwin (1871) distinguished between primary and secondary sexual characteristics and considered the organs of reproduction the primary ones. The secondary dimorphisms are the result of sexual selection. Thus as a result of intrasexual selection pressures, males have been selected for characters such as large body size, weapons and aggressiveness which are valuable in assisting them in competition for females. There is no advantage for females to possess such features. In fact, such features contribute to shorter mortality among males. Daly & Wilson (1983) suggested that natural selection often promotes early death by favouring types whose energetic and dangerous life-styles can lead to an early grave. The very qualities that permit males to compete for mating opportunities increase their risks. Song and bright colours make male birds more conspicuous to predators as well as to their potential mates. In contrast, females of such species often do not sing and are drab in colour. In mammals, females tend to lack weapons and to be smaller and less aggressive than males.

Intersexual selection pressures are another source responsible for sexual dimorphisms of males and females. We have discussed the difficulties in making a clear distinction between the two sources of sexual selection pressures but there are instances where the distinction is clear. The classic example involves male ornaments in birds such as long bright tail feathers to which I alluded earlier. Darwin (1871) proposed that even though they might be detrimental to the survival of the male, such characters evolved to attract females. Male birds such as the peacock and the bird of paradise may have evolved long pretty tails that attract females but also attract the attention of predators and hinder their flight. However, the gains derived from being more attractive to females outweigh the possible loss arising from a trait that increases the chances of being caught by a predator. Males that do not attract females would not leave their characteristics in the gene pool of the species. In addition, Zahavi (1975) and Zahavi & Zahavi (1997) argued that if a male survived despite the handicap of an ornament, this would serve as a signal to the female that he must be exceptionally strong and healthy. Thus the tail serves as a signal of true information about his physical qualities. If growing a long tail costs the male by diverting resources that otherwise would go to building other more essential parts of his body or his muscles, then only males that are strong and healthy could afford to pay this cost (Andersson, 1982; Zahavi, 1975; Zahavi & Zahavi, 1997). The rest would grow shorter tails and females would be able to discriminate male quality on the basis of tail length.

The notion that there is female preference for males with exaggerated ornaments have been tested in a few carefully controlled experiments. Andersson (1982) studied the African widowbird, a polygynous species that nests in the grassland of Kenya. Females are drab, mottled brown with a 7 cm tail and males are jet black, with bright red epaulets and a 50 cm tail. The male attracts females to his territory by flying low above the grass with tail fanned out. The female builds her nest in the chosen male's territory and rears her young with no aid from the male. Andersson selected a set of males with established territories of similar quality and divided them into the following treatments. Group 1 had their tails shortened to 14 cm. Group 2 had extra feathers glued to their tails so as to extend it to 75 cm. Group 3 had their tails shortened and the feather reglued to their normal length thus serving as a control condition. Because male–male contests for territory are decided by the flaring of epaulets, there was no change in territory boundary arising from the manipulation of tail lengths nor decline in epaulet display. Interestingly enough, newly arriving females preferred the territories of the long-tailed males over those of their short-tailed rivals by 4 to 1 and the control group by 3 to 1. These results suggest that tail length does make a difference in the female's mate choice.

One problem with Andersson's (1982) study is that both male–male contests and female choice operate successively. Møller (1988) corrected this problem by capturing male barn swallows after they had set up their territories but before the females had arrived. Males have a distinctive V shape to their tails which are prominent in their airborne courtship display to females. In Møller's experiment some males had their tails shortened to 8 cm, while others had them lengthened to 12 cm, and the third group had their tails shortened and then the section reglued to the normal 10 cm length. The short-tailed males had to wait four times as long to attract a female as those with an enhanced tail. The controls were only at a twofold disadvantage. The long-tailed males also began to breed sooner than the others and were more likely to rear two clutches during the breeding season. However, studies by Smith & Montgomerie (1991) produced similar results but also uncovered the following. DNA fingerprinting used to identify the parentage of the offspring indicated that the long-tailed swallows lost about 40 per cent of egg fertilisations to other males. Although the long-tailed males may pair more quickly with females they do not father as many offspring as controls with shorter tails. They suggested that the artificially-lengthened tails may interfere with the male's ability to guard his mate. Such results indicate that although tail length is a factor influencing female preference, it may not result in enhanced reproductive success unless it provides an honest signal of his abilities. Perhaps male swallows with naturally elongated outer tail feathers might exhibit such abilities.

Hamilton & Zuk (1982) hypothesised that males with the 'showiest' ornaments are more resistant to parasite infection than those which are less 'showy', because males have been under selection pressure to advertise their hereditary resistance. They found that there was a positive correlation between chronic blood infection and male colouration and song complexity among 109 passerine bird species. More recently, Milinski (1997) argued that female sticklebacks judge males upon the basis of the redness of their belly, and that the depth of colour is an indicator of males' resistance to parasitic infection. Because the red pigment is physiologically costly to produce, it is an honest signal of health. However, Read & Harvey (1989) argued that there are difficulties with the Hamilton–Zuk hypothesis, and in a study of 113 species of passerine birds, found an inconsistent relationship beween brightness of the male feathers and the prevalence of blood parasites across species. Another study involving 10 species of birds of paradise which controlled for variables such as body size, diet and altitude, indicated that the brighter the males, the greater their parasite burden or load (Pruett-

Jones, Pruett-Jones & Jones, 1990). These researchers also found that the promiscuous species of birds of paradise which were brighter than the monogamous species, had a higher percentage of males that were host to at least one parasite and thus were more parasite prevalent.

Although the cross-species comparisons produced mixed results, within-species studies produced results that are generally favourable to the Hamilton–Zuk (1982) hypothesis. On the whole, the more flamboyant a male's ornaments, the lower his parasite burden, the more females favour the males with fewer parasites. Møller (1988, 1990) found that barn swallow males that are heavily parasitised by a blood-eating mite have shorter tails than parasite-free ones. Unmated males are more often and more heavily parasitised than mated males. This may explain the female swallows' preference for long-tailed males. Zuk *et al.* (1990) manipulated the parasitic load of male red jungle fowl by infecting them with an intestinal roundworm. This procedure reduced the ornamental chacteristics of these males relative to unparasitised males. Clayton (1990) produced similar effects on rock dove by infecting some with a parasite that reduces the males' courtship display relative to non-infected males. A recent study by Thompson *et al.* (1997) is the first to monitor the long-term effects of parasites, through marking and recapturing the birds, on colour, growth and plumage in individual birds. They studied finches and found that mites and viral infections during molt are correlated with poor physiological condition and reduced development of bright male plumage during the molt period. In general, these results indicate that male characters that are effective in enhancing female preference also signal their good health and resistance to disease.

As noted in our earlier discussion, sexual dimorphism is less prominent in monogamous than in polygynous species because of lower parental investment by polygynous males. Although there are ecological factors that result in polygyny in some species of bird as well as monogamy in a few mammalian species (Emlen & Oring, 1977), the ability of male birds to incubate the eggs and feed their offspring and the inability of most male mammals to care for their young are important considerations. Thus sexual selection is more intense among male mammals than birds and results in more prominent sexual dimorphism in the former than the latter. The complex interplay of intra- and intersexual selection pressures impinging upon males makes it difficult to disentangle the magnitude of their contribution to a particular species. Are larger and stronger males with striking ornamentation more attractive to females or are they more successful in their male–male contests? Probably both.

Extra-pair matings among monogamous birds

The preceding discussion was premised by the assumption that the monogamous partnerships of birds provide an ideal to which many people should aspire. However, recent and more careful studies have revealed that appearance can be deceptive and that many birds are far from models of fidelity. In the previous section of this chapter, we had examined the traditional view of monogamy in which each partner is regarded as benefitting from mutual cooperation. However, we are now aware that because of individual selection, animals should behave selfishly. The traditional view focused on the cheating males who attempt to inseminate the females of other males and hence parasitise their paternal care. By doing so, they could increase their reproductive success and do so at minimal expense because sperm are cheap. Females were regarded as likely to be coy, since they need only one successful copulation to ensure that their eggs are fertilised. It was argued that infidelity was costly for a female since her partner would retaliate by reducing his investment in her offspring. However, it can go even further.

With ring doves, courtship and incubation behaviour in males is dependent on gonadal hormones. In addition, the male's behaviour provides the female with a means of predicting his level of parental investment. A male showing appreciable courtship is likely to display appreciable parental behaviour. However, males display less courtship and more aggression toward females that have been recently associated with other males than to females that have been isolated (Erickson & Zenone, 1976). Females in the former condition displayed immediately in the presence of the male in contrast to the coyness of those in the latter condition. Zenone, Sims & Erickson (1979) found that when a male ring dove discovers his mate with another male, he often attacks her. These findings have been interpreted as reflecting an adaptation that reduces the chances of cuckoldry.

A recent review by Birkhead (1996) has provided material that questioned the notion that birds form strong pair-bonds and remain faithful. Field studies have indicated that extra-pair copulations do take place and that paired males also protect their mates from the sexual advances of other males. If the total blame of such infidelity is among males, we would not be surprised. As a result of the technique of DNA fingerprinting, which allows for the unambiguous assessment of parentage, there is evidence that females also engage in multiple matings in the wild (Birkhead *et al.*, 1990). This surprising outcome raises the question of why females engage in extra-pair copulations? We would not have expected this according to the earlier models. Is there any adaptive significance for females who copulate with more than one male during a reproductive cycle? Such a functional question may best be answered by analysing the mechanisms of underlying physiological processes. This leads us to a discussion of the mechanism of sperm competition, a topic that will be brought up again in the following section on sexual selection in primates.

In many species, the second of two males to copulate with a female typically fertilises most of her eggs. This phenomenon is known as 'second or last-male precedence' and is seen in insects, crustaceans and birds. Birkhead, Clarkson & Zann (1988) performed two experiments with the zebra finch, both designed to mimic situations that take place in their natural setting. The first experiment involved mate-switching in which two males replaced each other. Despite the fact that both responded with a similar number of copulations, the second male fertilised more (75%) eggs than the first. Thus the 'last-male effect' was documented in the laboratory setting. The second experiment was designed such that the first male secured nine copulations on the average, but the second male was allowed one insemination. The latter fertilised over 50 per cent of the eggs and demonstrated the potency of a single extra-pair copulation. Subsequent analyses indicate two factors responsible for this effect, the timing of insemination relative to egg-laying and the quality of the sperm. The relative numbers of sperm from different males that are in the female tract at the time of fertilisation determines the outcome of sperm competition. Because of the nature of the bird reproductive system, inseminations close to egg-laying are less successful. By the time the second insemination took place, some sperm from the first insemination had already died or had been lost from the female tract.

Prior to the findings reviewed by Birkhead (1996), it had been assumed that sperm competition is driven primarily by selection on males. They safeguard their paternity and fertilise other females and thereby increase their reproductive output. In contrast, the only benefit monogamous females gain from copulating with more than one male would be an increase in the quality of her offspring. Birkhead's work questions the view that selection operates more intensively on males than females in terms of sperm competition. He suggested that females that have mated with several males may also have a physiological ability to favour the sperm of one male over another's. Since fertilisation takes place inside the female's body, selection may act to give them some control over paternity. Birkhead estimated that 'over 99% of the sperm a male inseminates is rejected by the female'. The physiological ability to determine paternity in this manner is referred to as cryptic female choice. This concept refers to 'female processes that affect male reproductive success and occur after the male has succeeded in coupling his genitialia with those of a female' (Eberhard, 1996). From Birkhead's perspective, the female is believed to have the mechanism to modify the degree of rejection slightly one way or the other and thereby change the odds for the competing sperm.

Given that females gain more from monogamy than males why should a female bother to engage in extra-pair matings? One suggestion is that females prefer higher quality males than her mate. This then raises the question of why the female didn't choose the highest-quality male in the first place? The answer is that this may not always be possible. Among migratory birds such as swallows, males generally arrive at the breeding grounds a few days before the females. The first females to arrive have the choice of the highest quality males but later-returning females have a more restricted range of males, and of lower quality. Rather than forgo breeding during that season the female may choose a lower quality male. However, she doesn't completely forgo the chance for some higher quality offspring and hence engages in extra-pair mating. What constitutes high quality when female birds choose their mates? It varies among species. For example, female zebra finch prefer to mate with males with phenotypic traits such as a high song rate and a red beak. It is not clear how they benefit from this behavioural choice of mating partner. In another species of bird, the swallow, females that are paired with males with short and asymmetric tails often seek extra-pair matings with males with longer and more symmetric tails. In the previous section, we encountered the experiments by Andersson (1982) and Møller (1990) that demonstrate the advantage that male swallows with longer tails (whether artificial or real) had over their shorter tail competitors.

The recent discoveries of female infidelity among some species of birds suggest that a distinction be made between genetic and social monogamy. Although birds may pair-bond, not all of the young in the nest are necessarily sired by the same male. Undoubtedly, there are many benefits for both the male and female to form a pair-bond; otherwise male birds would adopt the typically mammalian polygynous mating strategy. And of course, the young benefits more from the monogamous than the polygnous mating system.

Molecular studies of paternity have indicated that sperm competition takes place in mammals, including species such as humans regarded as either monogamous (Birkhead, 1996; Van den Berghe, 1975) or polygynous (Short, 1979). Birkhead estimated that between 1 in 10 to 1 in 20 offspring are the result of extra-marital matings. However, the mechanisms mediating fertilisation success in birds are fundamentally different from those in mammals. There is no consistent 'last-male effect' in mammals because the fertilising lifespan of mammalian sperm is short. It has to undergo a physiological preparation for fertilisation known as capacitation. Unlike the situation with birds, the mammalian eggs remain fertilisable for only a short time. A male must time his copulations so that that his sperm undergoes capacitation just at the time the female ovulates. Thus, there is little biological advantage in males or females to engage in extra-pair copulations after the female has ovulated. The pair is highly likely to have copulated during the peak of the female's oestrous period. Yet extra-marital matings occur among humans, partly because of the operation of a proximate causal factor such as 'pleasure' attained during the act but perhaps there may be an ultimate causal factor operating as well. Perusse's (1993) survey and analysis, which is fully discussed under 'Human Social Status and Reproductive Success' may be relevant to this issue. However, his analysis focused on the benefits to males and had little to say about females.

SOMATIC AND GENITAL SEXUAL SELECTION

Our discussion on sexual selection has focused on sexual dimorphisms relating to external physical features such as size, natural weapons, physical ornaments, fur or feather colour, and the like. This is referred to as 'somatic selection' (Short, 1979). As previously discussed there are difficulties in partitioning the contributions of male competition and female attraction to the evolution of such characters. In his studies comparing the mating systems and dimorphisms of four species of primates, Short extended his analysis to the genitalia. This led him to study gonadal and genital development in humans and the great apes such as gorilla and chimpanzee found in Africa and orangutan from Borneo. The results showed that somatic size and genital development are not necessarily related to one another. Male gorillas weigh around 250 kg and are twice the size of females. These apes have tiny 10 g testes in contrast to the large 60 g gonad of the smaller 50 kg male chimpanzee who is only slightly larger than the female. This apparent paradox was explained by Short in his discussion of the different mating systems of these species and their relationship with somatic and genital development.

Gorillas are much larger animals than chimpanzees (as well as humans) and have a polygynous mating system in which one male absolutely monopolises access to several females. We have discussed the intense selection pressures faced by males of polygynous species and their effects on somatic selection. The social unit of gorillas consists of a dominant male, who attains his status through inter-male competition, and a number of females. Because the male has sole access to the females in this unit, there is only a slight risk that a fertilisable female will receive ejaculates from any other male. Delivery of large quantities of sperm to a mate, therefore, carries little advantage for male gorillas. A comparable situation applies to the other large ape, the orangutan. Because they are arboreal, fruit-eaters and dispersed over large areas, orangutans do not form clearly defined units. However, when the females are in oestrous, there is male competition for access to her, and the largest dominant male usually becomes her mate.

In polygynous species, where one male has exclusive access to several females, intense male–male competition results in the selection for larger and stronger males resulting in the pronounced size differences between males and females. The gorilla and organgutan provide an interesting illustration of these polygynous attributes. The dominant male may have a harem of a few females but each of them will come into oestrous once or twice every three or four years or so. During the rest of the time she would either be pregnant or lactating and thus be anoestrous. For these reasons the male may only have the opportunity to copulate only once or twice a year, as mating is limited to the brief period of time when the female is in oestrous. Such considerations explain the relatively small testes of males from these species, the high proportion of morphologically abnormal spermatozoa in their ejaculate, and the absence of any pronounced sexual swelling in the peri-anal region of the female at the time of her oestrous. Females of polygynous species do not need to develop pronounced sexual swellings to advertise their state, as the dominant male constantly attends her.

The situation of the smaller ape, the chimpanzee, is very different. It is a medium-size group-living primate that has been selected for sociality as a consequence of the benefits of group foraging. They feed primarily on fruits and the group size varies with the availability of fruit-bearing trees. The troop size can be as large as 50 or so if there is a large fig-bearing tree in the vicinity. Their mating system has been classified as 'promiscuous' or multi-male mating (Short, 1979). Inter-male competition for access to females still results in males being larger than females, but somatic dimorphism is not as pronounced as that of the large apes. In fact size dimorphism is slight because in most troops, most adult males interact with oestrous females. The fact that several males mate with a female when she is in oestrous introduces the element of sperm competition. The male who deposits the greatest number of viable spermatozoa in the female's reproductive tract will enhance his chances of producing an offspring. This situation results in strong selection pressure for increased sperm quality and capacity, thus resulting in increased testicular size. This attribute will also be favoured for males exhibiting high copulatory rates with the female members in the troop during their prolonged period of oestrous.

From the female chimpanzee's perspective, if she is going to mate with several males, it would be advantageous to advertise her state of oestrous. This results in the development of pronounced sexual swellings at the time of ovulation that can be seen at a distance by the males. Because the dominance structure of these animals is less rigid than it is among the other apes (Goodall, 1986), females mate freely when they come into oestrous.

SEXUAL SELECTION IN HUMANS

What is the relevance of Short's (1979) analysis to the understanding of humans? The fact that men are 15 to 20 per cent bigger than women indicates that size dimorphism is slight but nevertheless evident. Short pointed out that monogamous species that pair for life exhibit little or no somatic dimorphism in overall body size. Marmoset monkeys and gibbons were cited as good examples of monogamous primates that document this point. Upon this basis Short suggested that humans are not inherently monogamous. However, the relatively small size (compared to the chimpanzee) of human testes which weigh around 20 g, the high proportion of morphologically abnormal spermatozoa, and the lack of cyclical sexual swelling in the ovulating female suggests that humans are not adapted to the multi-male promiscuous mating system evident in chimpanzees. The fact that human females do not exhibit visible displays indicating imminent ovulation is interesting because most female primates do. The human mid-cycle appears to be cryptic, although there are signs of changes in bodily odours and enhancement of specific sensory modalities (Doty *et al.*, 1981; Robinson & Short, 1977). This phenomenon has been labelled as 'loss of oestrous' or 'concealment of ovulation' and is a result of human evolution. These labels refer to both the absence of conspicuous external signs at mid-cycle and the human tendency to engage in sexual behaviour more or less throughout the menstrual cycle. What are the consequences of such a tendency? Some theorists (e.g. Lovejoy, 1981) have suggested that it may reflect a mechanism for encouraging male fidelity.

In common with some primates, but unlike other mammals, humans exhibit a menstrual cycle, rather than the oestrous cycle of sexual receptivity. The menstrual cycle is characterised by periodic bleeding due to sloughing off of the uterine lining. Animals with menstrual cycles are sexually receptive most of the time and this contrasts with the oestrous cycle where an animal is receptive only at the time of ovulation. Lovejoy (1981) has proposed that continual sexual receptivity helps to maintain the interest of the male. Because the time of ovulation is concealed in women, a man faces the situation where he must copulate regularly with the same woman if he wishes to ensure fertilisation. He also faces the problem of guarding his mate against advances from other males to ensure certainty of paternity. The low and uncertain chance that a particular mating will lead to fertilisation not only encourages continual male attentiveness but also reduces the benefits of opportunistic copulation with other women. Besides, a man tempted with such a possibility will run the risk of attack from another male who may be the other woman's mate (Smith, 1984).

How would we explain human testicular size relative to that of other primates? Certainly the male gonads of humans is much smaller than that of the chimpanzee, though much larger than that of the other large apes. We noted that genital selection is influenced by the mating system, and is ultimately a reflection of the frequency of copulation and the number of sexual partners. The low copulatory frequency of gorillas and orangutans places little selection pressure for increased sperm production, and explains the very small testes of these apes. This contrasts with the enormous gonads of male chimpanzees whose behaviour is characterised as a multi-male mating system. The copulatory frequency of humans is intermediate between those of the great apes and the chimpazee and the size of man's testes reflects this relationship. Short has also suggested that social bonding rather than reproduction seems to be the reason for the vast majority of human acts of intercourse. Because the normal frequency of copulation among humans is much less than that of the multi-male mating chimpanzees, we would not expect as high a selection pressure for increased testicular size in humans. Short's (1979) depiction of size dimorphism and differences in gonadal/genital development amongst male and female primates of these species are shown in Figs. 2.4 and 2.5.

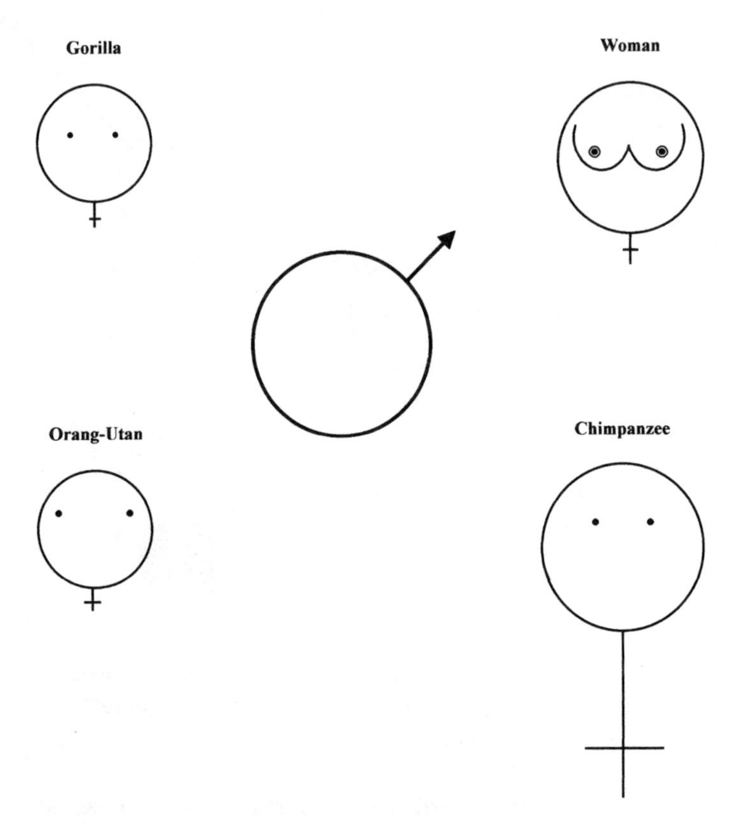

Fig. 2.4 The male's view of the the female, illustrating the relative sizes of the nulliparous breast and the external genitalia. (Redrawn from Short, 1979, copyright 1979, with permission from Academic Press.)

Pair-bonding in humans

Although there are persistent incidences of extramarital matings and some polygynous societies, humans tend to be predominantly monogamous. In most societies, marital relationships are sufficiently stable that men can invest in their mates' children with reasonable confidence of paternity. Daly & Wilson (1983) stated that although most marriages are monogamous, the variance in child production is still likely to be somewhat greater for men than for women'. Short (1994) viewed such events as indications that we are basically a polygynous primate in which the polygyny usually takes the form of serial monogamy. Because men have a longer fertile life than woman, a man is likely to produce more offspring than a woman. If older men remarry, they usually do so

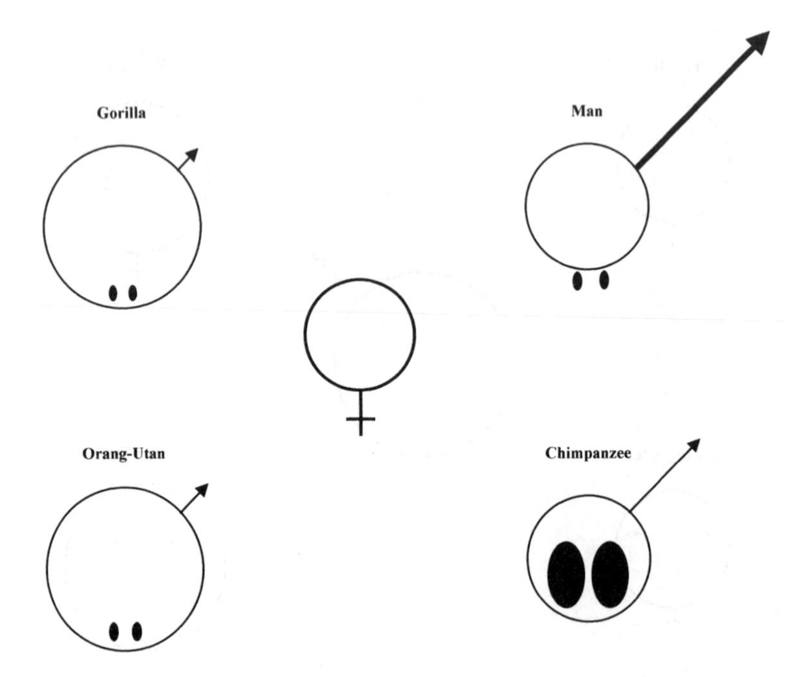

Fig. 2.5 The female's view of the male, illustrating the relative sizes of the penis and the testes. (Redrawn from Short, 1979, copyright 1979, with permission from Academic Press.)

with younger women. Nevertheless, Daly and Wilson commented that humans have established relatively stable mateships that persist even though they engage in highly personalised social interactions with a large number of others. McFarland (1993) has argued that the frequency of strict monogamy among humans is difficult to evaluate and he emphasised that it is the individual's total reproductive activities rather than those within a mateship that is of evolutionary importance. However, he acknowledged that some features of human relationship are greatly facilitated through monogamy.

The reproductive strategy characteristic of humans involve high investment in each partner for a few offspring relative to those of most other mammals and primates. Human infants are born at intervals of about two years. Reproductive maturity occurs late, and a woman usually gives birth to few offspring in her lifetime. These offspring require considerable parental care if the strategy is to be successful. Human mothers experience considerable difficulties in raising children without help. This is because compared with other primates, human infants are extremely helpless (Passingham, 1982). The helplessness of human infants arises from their immature brains. The human brain is four times as large as that of a non-human primate of equivalent stature. Although the size of a baby's brain at birth is very large compared with the size of its body, the baby is only partly functional. A human baby takes twice as long as a gorilla or chimpanzee to reach the point where it can support its body with its limbs. We have already discussed the possible utility of 'concealed or hidden oestrous' as a mechanism enhancing male fidelity and thus monogamy.

To say that humans tend to pair-bond does not imply that they are naturally monogamists. Once established, a pair-bond can be broken but usually with trauma. Yet attempts by some men to establish a promiscuous pattern of sexual relations seem to be very unsatisfying. There seems to be a deep human need for lasting relationships as opposed to fleeting sexual encounters. Humans tend to be picky about their sexual partners, and tend to stick with one partner when they find a compatible one. We also differ from other mammals, including other primates, in another respect. In most mammals, adult males are ready to mate almost any time while females are receptive only when in oestrous. This phenomenon has implications for an understanding of human mating preferences. The cultural expression of the human tendency to pair-bond is marriage which is a social institution that gives moral and legal sanction to this biological propensity. Although sexual intercourse does take place outside of marriage, in most societies, most stably mated pairs are married. So why is marriage and pair-bonding so universal in humans? The suggestion is that they contribute to the survival of the offspring and hence the fitness of both mates.

From the preceding discussion we conclude that the family is the cornerstone of the social order of humans. Although human mating patterns vary among societies, it is likely that humans have become progressively more monogamous in response to an increasing requirement of parental care. We noted that humans have the longest period of development from birth to sexual maturity among the primates. It would be very difficult for infant humans to be successfully brought to maturity without close cooperation between the parents. In most societies, the father stays with the mother when the children are young. He usually has more than one child by a particular woman, and he usually helps to support her and to care for their children. Although men have a vested interest in the welfare of their children, they gain from promiscuity, especially if their children by another woman will be cared for by another man. Or course, there will be forces that counteract the prevalence of such adultery. For this reason many societies attach great importance to the establishment of paternity. A husband who has parental duties has more to lose than a woman does if she has a child by another man. The woman's genetic relationship to the child is not in doubt, so any maternal care she displays will benefit her own genes. A man does not have this certainty so he usually guards his mate from other men. Recent data on male and female differences in criteria for mate choice document this point.

Human mate choice

Our discussion on human sex differences has presented many intriguing, informative, though provocative ideas based on deductions from evolutionary theory was well as data from non-human animals. Such ideas have attracted the attention of a new generation of social scientists working within the domain of evolutionary psychology and personality/social psychology. These researchers have designed and conducted psychological experiments and tests on humans that are relevant to the notions that we have been examining in the previous sections of this chapter. The pioneering cross-cultural psychological survey conducted by Buss (1989) is an excellent example of such research. In his massive study, Buss collected data from respondents from 37 samples drawn from 33 different countries located on six continents and five islands ($N = 10\,000$). The mean age of the respondents ranged from 17–29 years and consisted of men and women who were asked to respond to questions collected by native residents who were part of Buss's research team. The subjects were asked to provide biographical information such as the age in which they would prefer to marry, as well as the ideal age difference between themselves and their future spouse.

The participants in Buss's (1989) study were also asked to respond to a psychological questionnaire in which they were presented with a list of 18 psychological and social characteristics. Examples of these items include 'dependable character', 'sociability', 'chastity', 'intelligence' and other attributes about a person. The subjects were then asked to rate each of these attributes on a four-point scale for their importance in choosing a mate. Most of these attributes were of general social and personality variables that were not directly relevant to evolutionary conceptions. However, there were some 'target variables' that were interspersed in this general list of attributes and Buss's analysis focused on the former. These target variables pertain to cues related to resource acquisition and to characteristics signalling reproductive capacity of a prospective mate.

In our earlier discussion of mate choice among non-human animals, we observed that females are attracted to males that control territories or physical resources that benefit the female and her future offspring. An analogous variable in human society would be the 'financial prospects' of a potential mate. The results of Buss's study indicated in 36 out of the 37 cultures sampled, women valued this attribute in a potential mate more than men did. This finding supports Buss's prediction that women should value more than men attributes in a potential mate such as ambition, industriousness and earning capacity. In 29 out of the 37 samples, women valued the ambition and industriousness of a potential mate. A detailed examination of a sample that showed the opposite sex difference, where men rather than the women valued this attribute in a mate, may be explained in terms of specific local conditions. The Zulus who live in South Africa, when under the apartheid regime clearly expressed this preference. During this period of time, there was a division of labour in which Zulu women were the ones who built the houses, fetched water and performed other physical tasks. Because of the shortage of work for men in the villages and townships, the men often had to move to urban areas. This reversal of valued attributes probably, therefore, reflected the proximate local social and political pressures that were prevalent during that period in history. Assuming that recent changes in the political situation may change the economic prospects of the Zulus, it will be interesting to study their effects on male and female criteria for mate choice.

Buss (1989) also assessed the subjects' preference for the age of their ideal mate. In all of his samples, men showed preference for women who are younger. In general, they preferred having a mate who would be about 2.66 years younger. Conversely, women preferred to be married to older men who would be about 3.42 years their senior. Following these findings of self-reported preferences, Buss did a validity check and found that across the 27 countries, the actual difference between men and women at marriage ranged from 2.17 (Ireland) to 4.92 years (Greece), all showing the average age of wives to be lower than that of the husbands.

On the questionnaire the respondents were asked to rate the importance of 'good looks' in their ideal mate. All 37 samples indicated sex differences in the predicted direction – that men value this variable more than women. These results along with those on preferred age suggest that men value physical attractiveness and relative youth in potential mates more than do women. These results show generality across cultures and suggest male preference for females whose physical characteristics predict potential fertility. Alternatively, 'good looks' may be a signal of disease resistance. In a study of 29 cultures, Gangestad & Buss (1993) found a relation between the degree of pathogen level and the societal value of physical attractiveness analogous to long, beautiful tail feathers in male peacocks. They hypothesised that such 'good looks' serve as an indication that one was free of disease-causing microorganisms which were assumed to detract from physical attractiveness. This recent study raises the issue that physical high attractiveness in a mate should be as important a factor among women as among men because it is a signal of disease resistance.

A variable that is of evolutionary as well as of psychological interest concerns the respondents' rating of the importance of chastity or lack of previous experience in sexual intercourse, in choosing one's mate. An evolution-based prediction would be that males value chastity in potential mates more highly than females. This prediction is based on paternity probability; the husband is less able than the wife to be certain about the paternity of a child. The subjects' response to this item was less uniform than to the items concerning the importance of age and physical attractiveness and seems highly dependent upon their cultural background. Samples from 23 cultures including those of Africa, Asia, Middle East, South America and Eastern Europe support the evolution-based prediction. However, samples from countries in Western Europe such as Sweden, Norway, Finland, the Netherlands, France and the former West Germany indicated that prior sexual experience was not regarded as an important factor in a potential mate. Buss interpreted these results to indicate only moderate support for the evolution-based hypothesis. He conceded the influence of proximate cultural influences on the degree of importance placed on chastity. He also suggested that this variable is less directly observable than others assessed in the survey and argued that sexual selection for this character to be less specific than that for mate-guarding.

An alternative to Buss's (1989) explanation of the chastity results is found in the comparison between the social support systems common to the European countries where chastity is irrelevant as a criteria to those prevalent in 'old world' cultures that value this attribute in a woman. The possibility of raising another man's child poses less of a burden for those living under a social system where health care and education is provided by the state. In many western European countries, there are extensive day-care and pre-school facilities for children as well as state-supported education often to the level of university. Because 'parental care' is distributed collectively among members of the population, individual males are less financially burdened by the possibility of marrying a woman who because of past sexual history, may be bearing someone else's baby. In contrast, males from countries where child care is the sole responsibility of the family, and thus the father, incur a greater cost by marrying a woman whose sexual history is suspect. Thus, we would expect men from these cultures to accept only virgins as their brides. The prospect of supporting another person's child would be unacceptable. This analysis does not conflict with one that is evolution-based. Evolutionary theorists have assumed that behaviours influenced by genetic predispositions are modifiable by features of the physical and cultural environment (Lumsden & Wilson, 1981). (See Crawford & Anderson's (1989) discussion of selection for a facultative strategy in which certain environments lead to certain behavioural strategies, and other environments lead to a different strategy.) Thus, decisions and behaviours are contingent on current conditions including those from the social environment. There is less 'cost' for males in the western European countries than those from the old world if the child is not theirs. However, male sexual jealousy is prevalent in some of these cultures. For example, although the German and Dutch cultures have relaxed attidues towards sexuality (including extra marital sex), German and Dutch men are still upset when their partner shows sexual interest in a third person (Buunk *et al.*, 1996).

The results of Buss's (1989) ambitious project confirmed his predictions concerning earning potential, relative youth and physical attractiveness across cultures. The predictions regarding ambition–industriousness and chasity received moderate support. Cultural and ecological factors limit the universality of the evolution-based predictions about male–female differences on these variables. Nevertheless, the general pattern of results indicating that human males and females differ in reproductive strategies and that mate preferences represent important components of this strategy. The greater female preference for males displaying cues to high resource potential and the greater male preference for mates displaying cues to high reproductive capacity may represent adaptations to sex-differentiated reproductive constraints in the evolutionary past of humans. From this perspective, Kenrick & Keefe (1992) predicted that males' preference for relatively younger females should be minimal during early mating years, but should be more pronounced as the male gets older. They also predicted that young females would prefer somewhat older males during their early years but not have such a preference as they age. Because the subjects in Buss's study ranged from 17–29 years, it would be possible to test Kenrick and

Keefe's predictions from the former's data. This led more detailed studies on this issue by these authors.

Kenrick and Keefe analysed a sample of 218 personal avertisements from a 'singles' newspaper in Arizona that specified the age of the advertiser and a minimum desired for a partner. The results indicated that females tended to be fairly constant in their specifications throughout the age range. In contrast, males changed their preferences in a systematic fashion as their own age increased. In an attempt to relate such results from a limited source of information to the larger population, the researchers examined age differences in actual mate choice, as reflected in marriage statistics. They sampled all of the marriages occurring in two western American cities – Seattle and Phoenix – during January 1986. The results were consistent with their evolution-based hypothesis and those from their 'lonely hearts' study. The age of a man's wife tended to decrease in relation to his own age as he got older. Women usually married men who were just a little older than they were. Such results were also obtained when the researchers sampled 100 marriages recorded in 1923 in Phoenix, thus documenting consistency across generations despite changes in sex-role norms. Similar results were obtained from other cultures, such as those in Germany, Holland, India, and the Philippines.

The Kenrick & Keefe (1992) results are compatible with evolution-based models of mate selection. The changes in male preferences across the lifespan may be derived from the assumption that males would be interested in a female's fertility or reproductive value. Females would prefer males who are older possibly because they are more likely to build up economic resources as well as indirect resources such as social status. However, these men must be young enough to retain effectiveness as providers until the woman's last reproductive efforts have been reached. Social psychologists could explain these findings in terms of 'economic exchange' theory. They focus on proximate mechanisms and explain the age preference differences as a reflection of the means of maximising one's rewards in the exchange of asssets. Partners try to match themselves with others with similar social value. They assume that there is a conscious weighting of rewards and costs by the individuals. An evolution-based model bases the exchange process on the hard currency of biological fitness and reproductive value. These processes are not necessarily accessible to conscious calculation but reflect evolved adaptations that operate below the level of consciousness.

Human social status and reproductive success

We have seen the value and the power of evolution-based concepts derived from animal studies to the understanding of human mate preference. In the earlier sections of this chapter we had also considered data in animal studies that suggest a direct relationship between social status and male reproductive success. In humans, status defined in terms of resources, power or prestige has been subsumed under the notion of 'cultural success' (Irons, 1991). Studies of traditional societies during early and recent history indicate that cultural and reproductive success are correlated. When these variables were analysed in modern economically advanced industrial societies, the results indicated either an absence or an inverse relationship between socio-economic status and fertility (Vining, 1986). Such a finding raises problems for evolutionary theory. Tooby & Cosmides (1990b) have argued that adaptations are the results of selection operating in the past and that their relevance do not depend on whether or not modern behaviour is presently adaptive. If the current environment differs from the one during which most of the selection of proximate mechanisms for these attributes took place, it is not surprising that these attributes do not correlate with reproductive success in the present.

Perusse (1993) proposed that we identify specific features of the environment that may derail these proximate mechanisms. He suggested two features to account for the inverse relationship between male status and reproductive success in industrial societies. Contraceptive devices provide a means of separating sexual pleasure from its reproductive consequences, thus neutralising the influence of higher status on fitness. Social and legal enforcement of monogamy restrict variation in male sexual activity and may prevent higher status men from engaging in the practices of their counterparts in traditional societies. Perusse studied the French Canadians in Quebec to test his hypothesis that fertility is an inadequate measure of reproductive success in a modern industrial society. His study tested the adaptiveness of social status with actual mating and reproductive data. The subjects were 433 heterosexual Caucasian francophone Quebec men who were 25 years or older. They were either friends, relatives, fellow employees, etc. of the students from the two major French-speaking universities in the Montreal area. These students distributed the questionnaire to the respondents who returned them via pre-stamped and pre-addressed envelopes. The subjects were asked to respond to 132 items concerning personal matters. Information about their education, occupation and income were used to compute a composite measure of social status. Perruse computed two measures of male fertility. *Actual* or true fertility was based on the the total number of biological children fathered by the respondent and reflected mating in the absence of contraception. *Potential* fertility was assessed by the number of potential conceptions inferred through an equation relating number of sexual partners and number of coital acts per partner. The outcome of the analysis indicated that male social status was not predictive of actual fertility but was correlated with the number of potential conceptions that reflect the sort of contacts that would have contributed to reproductive success in a non-contracepting society.

There were a number of methodological problems concerning measures of status and fertility as well as sampling distribution in Perusse's (1993) study that were raised by the peer commentary following its publication. For example, Wilson (1993) argued 'striving to elevate oneself in the social hierarchy' is the relevant adaptation rather than 'attained status' as measured in Perusse's questionnaire. The latter reflects a composite of education, occupation and income (Bookstein, 1993) which describes the compromised outcome of many parties' behaviour. Despite these limitations, Perusse's results suggest that when conditions that characterise much of human evolution are made salient, the sexual motivation of modern humans may not differ significantly from those of their ancestors living in traditional societies. Another limitation of Perusse's analysis is that it did not consider the issue from the perspective of females. Birkhead's (1996) review, discussed earlier, suggested that extra-pair matings may also reflect an element of female preference for a 'higher-quality' male as potential parent of her offspring.

There is another issue that is critical to Perusse's (1993) emphasis on male status as a salient factor underlying reproductive success. Recent experimental data on captive pigtail macaque monkeys suggest that rank does not always predict the male's reproductive success (Gust *et al.*, 1996). The DNA profile-testing techniques used in the bird studies reviewed by Birkhead, were also used by these experimenters to assess paternity and the genetic representation of males in a captive group of 16 pigtail macaque monkeys. This method allowed Gust *et al.* (1996) to determine the relationship between dominance rank and reproductive success. The offspring were members of a group of 59 conspecifics. The results indicate no relationship between dominance rank and the number of offspring sired. The dominant male produced no offspring and the most successful male, siring eight, ranked fourth in the dominance hierarchy. These results may be valid only to monkeys of a specific species that live under unnatural conditions in captivity. Perhaps the relationship between high status and reproductive success may be positive if experimenters use DNA-identification among animals living under feral conditions, but as it stands, Gust *et al.*'s (1996) observations certainly raises questions about assumptions concerning this relationship. If their results were replicated in experiments with other mammalian species, we would be forced to reconsider the assumption that high male status invariably leads to greater reproductive success. In that respect, Perusse would be forced to develop a different explanation for the inverse relationship between status and reproductive success among men in industrial societies.

PROXIMATE FACTORS AND MALE–FEMALE DIFFERENCES

In the remaining sections in this chapter we will focus on proximate mechanisms of sex at the physiological and psychological level. We have examined the reasons why males and females are guided by different strategies but we have not studied what causes masculine and feminine behaviour patterns. We noted that male and female mammals differ at the chromosomal level. Males have the XY chromosomal pattern and females have the XX pattern. Note that the opposite is true of birds. Normally phenotypic sex corresponds to chromosomal sex but this relationship is not invariant. Early in its development the foetus is neither phenotypically male nor female. The internal sex organs exist in a form that can develop into either a male or female. This is due to the presence of both the Müllerian (which forms the female reproductive organs) and Wolffian (which forms the male reproductive organs) systems. Over time, one system develops and the other regresses. The male phenotype unfolds as a result of two processes. The Müllerian inhibiting substance actively suppresses the Müllerian system. Then testosterone is produced when the cells of the testes are in contact with a substance called the H-Y antigen. The latter is regulated by a gene located on the Y chromosome which, of course, exists only in males. This antigen then binds with a cytoplasmic receptor and the whole complex attaches to sites on the chromsomes. As a consequence of these events, testicular cells produce testosterone. The release of this hormone affects other cells, particularly those of the Wolffian system and result in further male typical development.

Females lack the Y chromosome and H-Y antigen and thus have little testosterone in their system. As a result of this, the Müllerian system continues its development and the Wolffian system regresses. No other additional events are necessary to produce a female. The critical factor for sexual differentiation is the presence or absence of testosterone during an early sensitive period in the organism's development. Even if the individual is chromosomally XX, the presence of testosterone by whatever means during the sensitive period would result in a male phenotype. Conversely, if the individual is chromosomally XY, the absence of testosterone during the sensitive period would result in the female phenotype. This principle was clearly demonstrated as a result of some elegant experiments performed on rats a few decades ago.

The choice of rats as an appropriate subject for effective experimental manipulation was propitious. Unlike most other mammals, the sensitive period for sexual differentiation in this species occurs shortly after birth (especially days 1–5). This fact makes it possible for an experimenter to directly intervene in the animal's sexual destiny without the complications of prenatal manipulations that would be necessary if, for example, primates were the subject of such experiments. The phenotype that was originally studied in great detail was the copulatory behaviour pattern exhibited in adulthood. Typically, female rats respond to stimulation by male rats by exhibiting lordosis, a copulatory posture where females plant their feet, arch their back, and avert their tail to the side. Males respond to the presence of a receptive female by mounting and intromitting (inserting the penis into the vagina).

Genetic (chromosomal) males that were castrated (i.e. had their testes removed) at birth (the beginning of the sensitive period) were tested when adults by the presence of an oestrous female. The males were given an injection of testosterone with a dose that was effective in activating male copulatory behaviour in non-castrated males. When tested, the early castrates failed to show the phenotypic sexual response. Instead, these animals showed female sexual phenotypic reactions when injected with oestrogens during the adult period. However, another group of males that were castrated when 14 days old (after the sensitive period) were responsive to the presence of an oestrous female following an injection of testosterone. These results indicate that males who are denied stimulation with testosterone during the sensitive period will be unable to manifest male phenotypic sexual behaviour. In contrast, males who are not castrated until after the sensitive period will not be impaired. They are able to function as males because of the presence of testosterone in their system during the sensitive period. Similar results were obtained in rats that were castrated during the sensitive period but who were given replacement doses of testosterone following the castration.

You may wonder if similar results would be obtained if genetic (XX) females were ovariectomised (removal of the ovaries) at birth. These animals were tested when adults with male 'studs' who had been primed with testosterone. Unlike their early gonadectomised male counterparts, ovariectomised females who were given oestrogen injections prior to the test responded to the male studs in a manner typical of a non-ovariectomised female. This finding indicates a 'feminine bias' in the brain even in the absence of ovaries. However, females who were ovarietomised at birth but who were given injections of testosterone during the sensitive period developed into phenotypic males. That is, when adults, they did not respond to oestrogen injections with sex-typical copulatory postures when in contact with male studs and instead, responded with male phenotypic mounting postures in the presence of female rats. Similar results were observed with non-ovariectomised females who were given testosterone treatment similar to the ovariectomised animals. However, females that were given testosterone injections after day 10 developed into phenotypic females. The sensitive period for hormonal influence on masculinisation had passed and the testosterone was ineffective.

There was a surprising outcome when young XX rats were given large doses of oestradiol (a specific form of oestrogen produced in the ovaries) during the sensitive period. These animals developed into phenotypic males and when tested as adults following the injections of testosterone, showed male sexual behaviour. They were unresponsive to the effects of oestrogen injections when adults. The reason for this paradoxical outcome lies in the fact that oestradiol, like testosterone, can influence gene expression. When testosterone enters cells, it is converted biochemically into several things, one of which is oestradiol (Callard, Petro & Ryan, 1978; McEwen, 1976). When oestradiol binds with special receptor molecules within the cell, the receptor hormone complex interacts with a segment of the cell's chromosomes. This influences the activity of some genes (O'Malley & Schrader, 1976). According to this analysis, normal masculinisation is achieved through the oestradiol derived from testosterone. This process in which testosterone is converted into oestradiol and the other androgen, dihydrotestosterone, is referred to as 'aromatisation' (McEwen, 1976).

If the above hypothesis were true, we might wonder, why aren't all females masculinised by their own or their mother's oestradiol? The answer lies in the production of a substance, alpha-fetoprotein, which is present only during the sensitive period and which attaches itself to oestradiol molecules in the bloodstream (McEwen, 1976). This action prevents oestradiol from entering cells in the critical brain regions and influencing gene expression. Because testosterone is not bound by alpha-fetoprotein, it is free to enter cells of the brain of males, where it is converted to oestradiol. The ability of alpha-fetoprotein to protect the brain of the XX female is limited because oestradiol is located in the extracellular region where it binds with alpha-fetoprotein and becomes inactivated. Because alpha-fetoprotein does not bind to testosterone, XX rats can be masculinised if given more oestradiol than their system can handle. The extra oestradiol will enter cells and cause masculinising effects. The production of alpha-fetoprotein occurs for a short period of time in rats and begins to disappear toward the end of the sensitive period (MacLusky & Naftolin, 1981). These results again document the principle that the sexuality of an organism is influenced by the presence or absence of chemical agents that affect gene expression during a sensitive period.

The sexually differentiated reactions described in the above experiments are the expression of a sexually differentiated brain (Harris & Levine, 1965). The sexual differentiation hypothesis explains why the removal of the testes from a day-old rat causes it to grow up endocrinologically and behaviourally as though it were female. The sex of his brain has been altered as a result of the absence of testosterone during the sensitive period. If a day-old female rat is injected with testosterone, it grows up as a male. This again demonstrates alteration of the sex of her brain. The male brain guides the development of male physiology and endocrinology and the female brain does the same for the female system. There are systems in the brain that control either male or female sex-typical behaviours. These behaviours extend beyond sexual or reproductive acts – the brain specified as male by testosterone determines the display of male emotional behaviour arising from aversion events. We will examine experiments that demonstrate how 'brain sex' influences many non-reproductive sex-typical behaviours.

THE CONCEPT OF 'BRAIN SEX'

The concept of 'brain sex' has been invoked to explain how early hormonal events influence the sex-typical reactions of an animal. Is there any empirical basis for this concept? There are clear differences in the anatomy of male and female brains particularly in the neurons in a subcortical region, the medial preoptic area (MPOA). The size of the nucleus of the cells within a region called the sexually dimorphic nucleus (SDN) in the MPOA of male rats is five times the size of that of females (Gorski *et al.*, 1978). This nucleus is first significantly different on the day of birth, and during the course of the first 10 days of postnatal life, the nucleus grows in the male but remains unchanged in size in the female. This sexual dimorphism is dependent upon exposure to testosterone or its oestrogen metabolite during the sensitive period. An injection of 1 mg of testosterone in five-day-old female pups results in increased MPOA nucleus volume when they are adults. Castration of the newborn male results in 50 per cent reduction in the nucleus volume when they are adults. Following the pioneering work of Gorski *et al.* (1978), there have been a flood of new studies implicating dimorphism in the anatomical development in the subcortical and spinal structures (Rand & Breedlove, 1987; Yahr, 1988) as well as development of areas in the cortex and hippocampus (Diamond, 1988; Juraska, 1991; Stewart & Kolb, 1988). These areas are involved in motivated activities other than those concerning mating, and their location and functional properties will be presented in subsequent chapters.

There is another way in which investigators can demonstrate that early hormonal effects influence 'brain sex' with sex-based behaviour arising as a consequence. This line of research indicates that early hormonal interventions can affect sex differences in non-reproductive activities. For example, there is sexual dimorphism in the behaviour of male and female rats when placed into a large novel apparatus and given what is known as 'the open field test'. During their initial encounter in this apparatus males tend to be inactive, show little exploration and reveal signs of fear through urination and defecation. In contrast, females tend to be more active than males and locomote about the perimeter of the apparatus as well as showing lower levels of urination and defecation. However, if five-day-old female rat pups are injected with testosterone, they will, when adults, behave more male-like in the open-field test than controls that did not experience early hormonal intervention during the sensitive period (Gray, Lean & Keynes, 1969). The 'treated' females showed less ambulation and more defecation than the untreated ones.

There is also sexual dimorphism in the behaviour of males and female rats on learning tasks. There is a well established finding that, when tested under food motivation in a multiple unit maze, male rats are consistently superior in spatial learning (Dawson, 1972; Tryon, 1931). Given this sex difference in spatial learning, one would expect a reversal of this difference through early hormonal intervention. Dawson (1972) examined the effect of neonatal feminisation of male rats on their behaviour in the Tolman sunburst maze and found that their scores were shifted in the direction of those attained by female rats. This subsequently led to a study where female rats were gonadectomised when pups and given testosterone injection during this early period (Dawson, Cheung, & Lau, 1975). This treatment elevated their spatial learning scores relative to those of control females. These experimenters also replicated the earlier report of the reversal of male's spatial skills through early castration or early oestrogen injection. In another study comparing oestrogen-feminised male rats with castrated controls, Binnie-Dawson, & Cheung (1982) concluded that sex-linked cognitive skills that are differentiated neonatally by same-sex hormones, can be reversed by opposite-sex hormone neonatal treatment.

Another example of sex-based performance differences is the finding that female rats acquire active avoidance responses more rapidly than males (Beatty, 1979; Denti & Epstein, 1972) but show poorer performance on passive avoidance tasks (Heinsbroek, van Haaren & van de Poll, 1988). Details on these procedures and mechanisms that mediate such behaviour are presented in Chapter 8 on the motivational effects of aversive stimuli on behaviour. On the basis of the preceding discussion, we would predict that neonatal treatment of female rats with testosterone would cause them to behave in adulthood like male rats when tested on active avoidance tasks. Beatty & Beatty (1970) found that a single injection of testosterone when the females were three days old, combined with testosterone treatment in adulthood at the time of testing, produces females whose avoidance behaviour resembles that of normal males. We would also predict that neonatal castration should improve the active avoidance behaviour of males. Although Scouten, Groteleuschen & Beatty (1975) found no significant differences in the performance of neonatally castrated male rats and controls, early castration combined with the prenatal exposure to cyprotene, an anti-androgen substance, enhances the avoidance acquisition and open-field scores. Males given this combination of treatment produce scores that are indistinguishable from those of normal female rats. This finding suggests that prenatal exposure to testosterone can result in male levels of active avoidance performance.

Female rats normally consume more sweet saccharin solutions than males (Valenstein, Kakolewski & Cox, 1967). By injecting testosterone in very young female rat pups, Wade & Zucker (1969) got them to show a male-like pattern of saccharin ingestion. The treated females also drank less saccharin when adults than control female rats. Such data along with those presented in the preceding paragraphs indicate that early hormonal manipulations have far-ranging effects on behaviour other than that of the copulatory sort. They substantiate the hypothesis that the presence of androgens (testosterone and dihydrotestosterone) during the critical stage in the growth of the nervous system will produce neural circuits capable of organising male sexual behaviour (Young, Goy & Phoenix, 1964). Although early androgens affect phallic (penis) development, which influence later male copulatory behaviour (Beach, Noble & Orndoff, 1969), they also masculinise the brain. Thus, male sex-typical behaviours not directly related to mating functions are also influenced by early hormonal events. This indicates that these events influence the biasing of 'brain sex' in addition to their obvious effects on priming and sensitising the growth of the phallus.

The sexual differentiation hypothesis and the concept of 'brain sex' is relevant to the analysis of the few studies done on primates. Male and female rhesus monkeys exhibit dimorphism in their play behaviour. In general, males engage in a greater degree of rough-and-tumble interactions with each other as well as with females when playing. Young male monkeys also initiate more play-fighting with their mothers than do females. Similar effects have been observed among young male and female rats. However, female monkeys develop masculinised external genitalia and exhibit masculinised social behaviour as youngsters if their mothers were given testosterone during pregnancy (Goy, 1970; Thornton & Goy, 1986). In contrast to rats, the rhesus monkey's sensitive period for sexual differentation occurs well before birth. This is also true for humans. Because of this, administration of testosterone to a newborn rhesus monkey or human baby would not have any masculinising effect. The general results of these studies suggest that organising actions of hormones on the central nervous system take place during periods of neuronal differentiation and growth. These actions can be thought of as predisposing individuals to respond to certain stimuli, both hormonal and environmental, in particular ways later in the animal's life (Stewart, 1988).

Ovarian hormones in sexual differentiation of the brain

Although the preceding discussion leads us to conclude that 'feminisation apppears to be the neutral condition' (Lisk & Suydham, 1967, p. 182) recent findings suggest that sexual differentiation of behaviour may also be influenced by ovarian hormones (Fitch & Denenberg, 1998). To make this case, Toran-Allerand (1984) proposed a distinction between the suppression of male or female attributes (demasculinisation and defeminisation, respectively) and the enhancement of male and female characteristics (masculinisation and femininisation), respectively. For example, certain hormonal manipulations can suppress female-typical behaviour without inducing male-typical behaviour (Yahr & Greene, 1992). In the previous section we noted that the SDN of the MPOA is larger in male rats and that increasing the size of the structure in females via early androgen treatment is intepreted as masculinising. Alternatively, such enlargement might be also be interpreted as de-feminising. Stewart & Cygan (1980) argued that while both ovarian and testicular hormones contribute to normal female and male development, their actions are not merely reciprocal and probably occur at different times in development. Research findings of the 1990s support this prophetic conjecture. As a result of their review of findings from many laboratories, Fitch & Denenberg (1998) concluded that ovarian hormones act during a sensitive period that extends beyond puberty, and perhaps to organise the brain of the female. They were especially interested in sex differences in the neural differentiation of the band connecting the right and left cerebral hemispheres, the corpus callosum.

The corpus callosum of male rats is larger than that of females (Berrebi *et al.*, 1988; Zimmerberg & Mickus, 1990). When the ovary of a female rat is removed as late as 16 days postnatal, her corpus callosum is larger during adulthood than that of a sham-operated female (Fitch *et al.*, 1991). However, treatment of low-dose oestradiol starting on day 25 prevents this increase. Callosal size is also increased by a combination of 'infantile handling' during infancy (see discussion of this topic in Chapter 7) and administering testosterone before day 8. In general, such results support the view that ovarian hormones play an important role in the development of the female brain and that the temporal parameters and mechanisms of 'ovarian feminisation' are different from those of 'androgen masculinisation'. It should be noted that although the corpus callosum may be involved in some non-reproductive behavioural sex differences, it does not directly regulate reproductive activities.

BEHAVIOUR AND NEURAL ORGANISATION IN CONTEXT

The discoveries described in the preceding section elicited much excitement in the field when they were first published. They conveyed the impression that such behaviour differences are determined solely by early exposure to hormones. When the results of a number of experiments revealed that the region of the brain where hormones exert their influence was one involved in the mediation of sexually dimorphic behaviour patterns, we might be tempted to think that the relation between hormone, brain and behaviour has been explained. Critics of this way of thinking cite the principle that behaviours occur in contexts, and that the probability of an act depends on the contextual and eliciting stimuli and prevailing organismic conditions (Birke & Sadler, 1985; Stewart, 1988). To document this principle, let us return to the issue of sexual dimorphism in play activity among monkeys and rats. Although play behaviour is greater among males and effects can be reversed through neonatal androgen injections to female rats and early castration of male rats (Meaney *et al.*, 1983), there are questions on how the effects are mediated. The hormonal treatment of young pups also alters maternal behaviour towards these pups.

Birke & Sadler (1985) conceded that hormonal treatments appear to act upon the developing rat brain but argued that these effects cannot be dissociated from possible effects on maternal behaviour. Mother rats nuzzle and lick their pups, in part because rat pups do not urinate or defecate without such tactile stimulation. You may recall my comments that organisation of many aspects of the rat's system develops after their birth. Male pups are licked more than females, apparently because of a difference in odour which is affected by the presence of testosterone in the male. Moore (1984) found that when maternal licking of male pups was experimentally reduced by rearing them with anosmic (loss of sense of smell) mothers, they show longer intermount intervals in their copulatory activities than males that receive normal amounts of licking. Providing female pups with tactile stimulation via a light brushing of the anogenital region increases the rate with which they perform intromission patterns when tested for masculine sexual behaviour when adults (Moore, 1985). Similarly, female rat pups injected with testosterone during days 2–14 and tested for later male sexual behaviour, have shorter intermount intervals than do untreated females. These results provide an excellent example of the importance of context in sexual organisation. A genetic mechanism influences the secretion of a hormone that affects the pups's odour, which influences maternal behaviour toward the pup, which in turn affects the pup's later behaviour (Moore, 1990). The effect of the hormone in provoking structural changes in the brain acts in concert with the effects of environmental variables, and thus modulates the development of behaviour patterns affected by these changed brain structures.

Studies of primate behaviour also document the importance of early social environment and context. The studies on social rearing or grouping of rhesus monkeys by Goldfoot *et al.* (1984) are particularly illuminating. When reared in mixed sex or heterosexual conditions, juvenile and adult male monkeys typically display a 'foot-mount clasp' with females while the latter 'present' themselves to the former with the female sexual posture. However, if male and female monkeys are raised when infants in an isosexual (same sex) group, the proportion of females showing the foot-mount clasps was much higher than in those in heterosexual groups. The larger and dominant females in the isosexual group 'foot-mount clasped' the smaller females. Amongst males, the frequency of presenting in isosexual groups was as high as that among females. The smaller males displayed the 'present' posture in the presence of the bigger and dominant males. These results demonstrate that sex differences in behaviour patterns usually considered to be sex-typical are determined by the social context.

SEXUAL DIFFERENTIATION IN HUMANS

We have noted that sexual differentiation in non-human primates is influenced by early hormonal factors. You may wonder if this principle is just as valid among humans. An examination of the literature on human sexual differentiation indicates that the evidence is mainly from clinical medical studies rather than from controlled laboratory experiments. For example, some patients experience a condition known as congenital adrenal hyperplasia (CAH) wherein genetic females are masculinised by a condition in which the adrenal glands produce large amounts of androgenic hormones (Hines & Kaufman, 1994). This condition is also known as the androgenital syndrome in which an XX individual has female internal sexual structures (ovaries) but masculinised external genitalia. The clitoris is so enlarged that that it may resemble a small penis. The degree of masculinisation depends upon how much testosterone was released and when its release occurred relative to the sensitive period. If the condition is detected at birth, adrenal androgen production can be controlled with lifelong cortisone treatment. This hormone is normally produced in the adrenal cortex but not among cases of CAH. These patients can also receive surgical procedures that feminise their partially masculinised external genitalia.

The study of CAH girls who had been detected and identified as girls is of particular interest. Although their cortisone therapy eliminated masculinisation soon after birth, these girls experienced prenatal exposure to androgens. The sensitive period for sexual differentiation in primates occurs during the last trimester of the prenatal period.

Ehrhardt & Baker (1974) compared 17 CAH girls with their unaffected sisters and found that they engaged in more rough and tumble play and in general, expressed less interest in clothes, dolls, jewellery, make-up and babies than their sisters. However, these CAH girls became sexually interested in boys when teenagers and the authors suggested that there was no reason to suspect that they were more likely to become lesbians. There is some suggestion that the foetal exposure to androgens contributes to the behavioural profile of some CAH females. This was supported by a study by Berenbaum & Snyder (1995) on the effects of prenatal androgens during foetal development on toy preferences and playmate preferences of boys and girls. Although prenatal androgen exposure did not influence playmate preference, it had an effect on toy preferences in the CAH girls. They spent more time playing with toys aimed at boys and less time playing with toys aimed at girls than the control group. More recently, Berenbaum (1999) found that CAH girls between the ages of 9 and 19 years showed sex-atypical interests in issues such as career choice and related activities than control girls.

Chromosomal females may be masculinised as a result of another clinical condition called *progestin-induced hermaphroditism* (Money & Mathews, 1982). Actually, these individuals are pseudohermaphrodites ('false' hermaphrodites) who have testes or ovaries, but not both. Unlike true hermaphrodites, their gonads match their chromosomal sex. Because of prenatal hormonal errors, their external genitalia are ambiguous or resemble those of the opposite sex. In the past, pregnant women who were prone to miscarriages were sometimes given synthethetic progesterone to facilitate anchoring of the zygote to the uterine wall. In addition to producing the desired effect, this hormone also has an androgenic effect. Because of the close chemical similarity between progesterone and testosterone, these steroid horomones are to some degree functionally interchangeable. The foetal daughters of women given this prenatal treatment were partially masculinised in a manner similar to the less severe cases of CAH. Such cases provide additional support for the notion that foetal androgens can predispose a genetic female to exhibit tomboyish behaviour during her adolescent period. However, this masculinising effect does not extend to sexual orientation.

There is also a condition, the *androgen-insensitive syndrome*, where genetic males produce testosterone but their cells either lack receptors for the hormones or the receptors are abnormal. With these cases, sexual differentiation continues in the female-typical pattern with female genitalia and female secondary characteristics. Yet this person has functioning testes that produce testosterone, but this individual has no ovaries or uterus and, of course, does not menstruate. Thus, this syndrome is a manifestation of a genetic defect that makes the person insensitive to androgens. Without androgens, the body follows the female pathway of development and sexual orientation.

A third type of pseudohermaphroditism is named the *Dominican Republic syndrome*, because it was first observed in a group of 18 boys in two rural villages in that country (Imperato-McGinley *et al.*, 1974). This syndrome is a genetic enzyme disorder that prevents the testosterone from masculinising the external genitalia. The enzyme 5-alpha-reductase facilitates the synthesis of dihydrotestosterone (DHT) from testosterone. A deficiency of this enzyme thus impedes the aspects of sexual differentiation that are normally under DHT influence. These boys were born with normal testes and internal reproductive organs, but their external genitals were malformed. Their scrotums were incompletely formed, their penises were stunted, and they also had partially formed vaginas. Because these children were regarded as girls at birth, they were reared as females. At puberty, their testes produced testosterone at levels normal in pubertal boys and thus, they experienced changes in voice, penile and testicular enlargement, and masculine patterns of musculature. Of the 18 boys who were reared as girls, 17 shifted easily into a male gender identity.

The Dominican cases provide an interesting example of the powerful influence of biology. If nurture were predominant, gender identity would be based on the gender in which the person is reared, regardless of biological abnormalities. However, these boys experienced pubertal biological changes that led to changes in gender identity and roles. It is likely that the pubertal surges of testosterone had activated brain structures that were masculinised during prenatal development. Prenatal testosterone levels in these boys were probably normal and could have affected the sexual differentiation of brain tissue. This happened even though the genetic defect prevented the hormone from masculinising the external genitalia. An alternative explanation would explain the ease of shift of gender identity because of the high value placed on being a male in Hispanic cultures. It may be easier for a child formerly regarded and reared as a girl to adapt to the change in status of that of a boy than it would for a boy to adapt to the social changes associated with being a girl. This is a hypothetical analysis, since there are no reported cases in which the latter condition has been observed.

In general, the examination and discussion of these clinical cases indicate the influence of biology on sexual differentiation in humans and the results are congruent with animal models and data. Certainly, social developmental variables play a significant role in determining an individual's gender role as well as the type of sexual activity expressed. But nevertheless, biology manifests its influence by the constraints which it imposes on the expression of the individual's sexuality. There is a single genetic locus, the 'testis determining factor' (TDF) locus, which provides a genetic switch (Clepet *et al.*, 1993; Goodfellow & Lovell-Badge, 1993) that determines sex. The TDF sets in motion a series of events that lead to masculine development as discussed in the section on 'brain Sex'. The presence of this gene located on the Y chromosome prompted Arnold (1996) to argue that 'nonhormonal factors may play a role in initiating sexually dimorphic neural development'. This switch organises the development of sexual dimorphism in a number of traits including body size and proportion, brain anatomy and physiology, and age at sexual maturity. However, constraints on the influence of this genetic locus may come about as a result of intervening events such as atypical social rearing conditions and excessive or minimal levels of hormones during the critical period for sexual differentiation. Such intervening events are relatively infrequent among most humans, and an individual is channelled along either a female or male pathway as a result of this locus. This notion is congruent with the sexual differentiation hypothesis which proposes that the presence or absence of testosterone during the critical period influences the direction of an individual's sexuality.

SUMMARY

The sexual strategies of males and females arise from the consequences of the nature of sexual reproduction and the inherent imbalance in male and female parental investment. The contributor of the scarce resource is the object of competition for those contributing less. The prevalence of competitive males and choosy females may be explained from this general principle.

An important proximate cause of male and female behavioural differences arises from hormones that influence morphological as well as psychological tendencies that express 'brain sex'. A critical factor for sexual differentation is the presence or absence of testosterone during an early sensitive period in the organism's development. However, sexual differentiation of the brain may also be influenced by ovarian hormones that play a role in the development of the female brain and consequent non-reproductive behavioural sex differences.

Although hormonal treatments appear to act upon the developing neonatal brain, their effects on the infant's behaviour cannot be dissociated from possible effects of the interaction of the mother with her offspring. A genetic mechanism that influences the secretion of a hormone which provokes structural change in the brain acts in concert with the effects of environmental variables. This modulates the development of behaviour patterns affected by these changed brain structures.

3

Parental/maternal activities

Offspring care or parental/maternal behaviour has received scant coverage in most books on motivation. This activity is manifested mainly by the mother, but in some species the father also engages in caregiving behaviour. Pryce (1992) defined maternal motivation as 'a female's tendency to make infants the goal of her behaviour where that behaviour can be described as promoting infant well-being'. This topic, if considered in motivation books, is often subsumed in the chapter on sexual activities and sometimes in the discussion on social attachment. Why should parental behaviour deserve more interest and attention? This activity ensures the survival of offspring and thereby enhances the parents' reproductive success. In most mammalian species active copulators who are unconcerned with the product or outcome of their mating activities may not leave many survivors to continue their genetic line. However, there is variation in the prevalence of parental behaviour among species. Wilson (1975, p. 168) declared that 'the pattern of parental care is a biological trait like any other; it is genetically programmed and varies from one species to the next. Whether any care is given in the first place and what kind and for how long are details that can distinguish species as surely as diagnostic anatomical traits used by taxonomists'.

The extent of parental care varies and is partly related to the complexity of the organism. Many aquatic invertebrates simply shed their eggs and sperm into the water and leave the embryos, arising from the consequent union, to fend for themselves. In contrast, caring for one's offspring is widespread among birds and mammals. Amongst the latter, there is a fundamental level of contact and interaction between the mothers and their young arising from the nursing relationship. In some mammalian species the contact between the generations ends abruptly following weaning, whilst in others, it persists long after. There are species that carry, feed, clean, protect, teach and discipline their offspring for many years. Care is costly and there must be a benefit that is manifested in the preservation of the individual's genotype and a relative increase in its frequency. In some species it may be beneficial to minimise parental investment in one offspring for the sake of producing many more offspring, whilst in others, the investment is maximised by excluding further productive effort. The former strategy reflects that of '*r*-selected' species (where *r* refers to the reproductive rate of the population) such as insects and fish that produce large numbers of offspring at frequent intervals, and the latter strategy reflects that of '*K*-selected' species (where *K* refers to the carrying capacity of the environment) such as humans and bears.

Amongst mammals, females often bear the brunt of parental care and other reproductive costs. An obvious reason is the nursing relationship alluded to earlier but there is also the factor of confidence of parentage. Trivers (1972) argued that females are guaranteed of their genetic relationship to their offspring, but males are rarely sure of their relationship to the offspring of their mate. There is always some likelihood that the eggs of the female could have been fertilised by another male. From this perspective, when the confidence of paternity is high, males provide extensive parental care. Such is the case with monogamous New World monkeys such as the cotton-top tamarin (Cleveland & Snowdon, 1984; McGrew, 1988; Tardif, Richter & Carson, 1984) and the marmoset (Box, 1977; Ingram, 1977). When confidence of paternity is low, males show little parental activity even though they may invest heavily in courting the female. The same pattern is evident in humans. In some societies where sexual promiscuity is common, men do not help their wives' children. Instead, they are fatherly to their sisters' children. For example, among the Navaho, the brother controls inheritance of his sister's children and he also takes on the paternal role of strict disciplinarian (Aberle, 1961). Kurland (1979) suggested that a man can be sure that his sister's children are related to him, but he has less confidence about his wife's children.

An alternative explanation for the prevalence of biparental care amongst monogamous animals entails economics. The common factor contributing to male parental care is *necessity* because of demands on the mother to secure resources for herself and her offspring. When a female can rear infants by herself, she allows little or no male involvement. Males are allowed to care for infants only when the demands on the female are so great that helpers are necessary. Tamarins and marmoset monkeys, often give birth to twins, and fathers are observed carrying and playing with the youngest offspring. This activity frees the mother to obtain food, which is necessary for converting to milk.

Although biparental care is usually found in monogamous species where successful rearing of offspring is dependent upon the work of both parents, this is not always the case. Males of some polygynous species participate in some aspects of parental care such as carrying, holding, grooming and protecting infants. An example of the latter is the reactions of male Barbary macaque monkeys to infants. In contrast to other Old World monkeys and mammals who rarely display parental interest, male macaques interact intensively with infants soon after birth (Lahari & Southwick, 1966). This well documented finding led Andreas, Kuester & Arnemann (1992) to use DNA fingerprint analysis to assess genetic paternity in their assessement of interactions of captive male Barbary macaque monkeys with 27 infants living in an outdoor colony. They found that males associate with infants as expected, but that fathers seldom participate in the social dyads, and that other males' preference for specific infants was not kinship based. These researchers concluded that their data cannot be interpreted within the framework of parental investment theory. However, these results may be restricted to monkeys living under conditions of captivity or that there may be specific life history variables that may be responsible for male proclivity for such behaviour. If future studies replicate similar results with other species living under feral conditions, the parental investment theory may have to be revised. We will revisit this issue in the section of this chapter dealing with human paternal care.

PARENTAL CARE AND ATTACHMENT

In mammals, parental care usually involves behavioural interactions between parents and offspring. One form of these interactions is that of parent and infant attachment. There has been considerable research on interactions between the mother and her young while male parental care has been less studied (Kleiman & Malcolm, 1981). A primary focus of research in mother–offspring relationship has been on the attachment or bond between the infant to its mother. There is diversity in the extent to which attachment develops amongst mammalian species. Much is dependent upon the species' life history and ecology. Until weaning, mammalian young are entirely dependent upon their mother as a source of nourishment as well as warmth, shelter and protection from predators. Because maternal resources are limited, mothers who care indiscriminately for their own and other neonates would risk depriving their offspring of essential nutrients while benefitting unrelated young. It follows that natural selection processes would favour the mechanisms that facilitate discriminative allocation of parental investment. This would be expressed in preferential treatment of the mother's biological offspring, which is made possible by the mother's ability to discriminate between its own and alien offspring. With these considerations in mind, Porter & Lévy (1995) predicted that the mechanisms for accurate recognition of offspring would most likely evolve in contexts in which mothers' own and alien young intermingle in the same area prior to weaning. Free-ranging ungulates such as goats and sheep whose offspring are precocial exhibit such recognition and differential maternal care.

Goats and sheep subsist on a variety of plant material and spend a large portion of the daylight hours grazing. The basic social unit is the female and her offspring, but closely related animals may also exist within the same home range. Newborn goats and lambs are highly precocial and single and twin births the most common (Gubernick, 1981a; Hersher, Richmond & Moore, 1963). The mother and her suckling offspring form large herds (goats) or flocks (sheep). Rather than remaining in a particular site when their mothers leave to graze, young kids and lambs often wander away and aggregate in peer groups. Mother and offspring regularly reunite, enabling the kids and lambs to feed.

When the females are about to give birth they show a tendency to seek isolation from the herd or flock. Immediately after birth the mother avidly licks her newborn, which is covered with amniotic fluid. Maternal licking continues until the offspring is dry and much of the birth fluids and membrane is ingested. Suckling usually occurs within the first two hours after birth. Maternal responsiveness is maximal at parturition but fades after a few hours in the absence of a newborn (Poindron *et al.*, 1980). Within this sensitive period a selective bond is formed between the mother and the newborn. The mother will suckle only her offspring and reject, often with threats and butts, any alien young exchanged for the mother's neonate. This characteristic of maternal selectivity lasts throughout the lactation period. Olfaction is the major sensory channel involved in the regulation of maternal behaviour, and olfactory cues are involved in the attractiveness of newborn during the sensitive period for maternal responsiveness as well as selective bonding and suckling.

Experimental studies on maternal selectivity in goats

The domestic goat is a member of the family Bovidae as are sheep. It lives in open habitats, found in herds and gives birth to one or two precocial young. Prior to parturition the doe withdraws from the herd, gives birth, and leaves her kids hidden for a few days. Although goat kids will initially attempt to approach and nurse from any mother, a mother will accept only her own kid (Collias, 1956). A short period of contact with her kid(s) after birth establishes a maternal bond (Hersher *et al.*, 1963), and following a three-hour separation, the mother will accept all her kids but reject alien kids (Klopfer, Adams & Klopfer, 1964). It was assumed that all the kids of her litter share some litter specific cues for recognition. The absence of post-partum contact with the young leads to the waning of maternal responsiveness and the rejection of all young, even after only a one-hour separation (Collias, 1956; Klopfer *et al.*, 1964). Chemoreception has been implicated in the establishment of maternal attachment, since olfactory impairment at parturition reduces subsequent own–alien-young discriminations (Klopfer & Gamble, 1966). These results led Gubernick (1981b) to propose that mothers in contact with their kids 'label' them and that such labelled kids are then rejected by other mothers. Gubernick hypothesised that a mother may label her kid directly through licking it or indirectly through the kid's milk intake. Licking is believed to transfer rumen microfauna to the kid's body surface, while ingestion and digestion of milk may influence body odours. Such a labelled kid may then be recognised and accepted by its own mother but be rejected by another mother. The labelling hypothesis was tested by an experiment where the mothers were given five minutes of contact with their own kid immediately post-partum. After a one-hour separation, the mothers were presented with their own and alien kids in a series of 10-minute tests. The kids were presented individually and in random order. One group of alien kids was kept with their own mother and another group was unlabelled as a result of separation from other goats and skim-milk feeding. The results indicate a low acceptance rate of labelled alien and high acceptance of labelled own kids. However, there was acceptance of two-thirds of the unlabelled kids by does who were not their mother. Gubernick explained this by hypothesising that an unlabelled kid might be acceptable to a mother within a few hours after parturition. The decision to accept or reject is based on the odour of the kid, but she gives it the benefit of doubt if there is no indication of a foreign label. After kids are four days old, mothers may recognise them through their vocalisation, since prior to this time, kids' vocalisations are acoustically similar (Lenhardt, 1977). With increasing age of the young, other cues become important to the mother in recognising her kids, especially from a distance.

Because it provides a simple and attractive explanation of maternal selectivity, the 'maternal labelling' hypothesis stimulated further research on this variable. Ronmeyer & Poindron (1992) obtained results which suggest that disparity in ages between the mother's own kids and the labelled alien young may have contributed to the differential rates of acceptance found by Gubernick (1980, 1981b). Furthermore, Ronmeyer *et al.* (1993) found that labelling of kids by their mother through licking and nursing does not determine their later rejection by an alien mother who has herself remained in constant contact with her own young. These researchers tested recently parturient goats for their responses to their own familiar offspring and two alien kids of the same age, including one that had been isolated since birth. In the first experiment all females accepted their own kid, but most (12/14) rejected both the labelled and the unlabelled alien. Thus, labelling of kids by their mother is not necessary for their later rejection by an alien female. In the second experiment the kids were placed into a wire-mesh cage at birth that prevented their mother from nursing or licking them. Although the cage manipulation disrupted maternal behaviour in a large proportion of the females, they were capable of recognising their own offspring and discriminating them from alien isolated (unlabelled) kids, even when there was no opportunity for maternal labelling. The experimenters suggest that when mothers and kids are left undisturbed after parturition, maternal labelling is not a necessary step for the establishment of maternal care.

One may wonder about the discrepancy between the results of Ronmeyer *et al.*'s (1993) study and the earlier ones by Gubernick (1980, 1981b). There were procedural differences in amount of initial exposure between the mother and her kid. In contrast to Gubernick's short exposure session of five minutes followed by a separation of one hour before the test, Ronmeyer *et al.* (1993) gave the animals four hours of exposure, and then tested the doe's immediate capacity to discriminate. Four hours of exposure might allow for the memorisation of subtle cues that may serve as individual signatures used for discrimination of the doe's own kid. There were also age differences in the test subjects in these experiments. The isolated (unlabelled) kids in Gubernick's experiments tended to be younger than mothered ones, and both types of aliens were older than the mother's own kids. In many respects, the conditions of the experiments by Ronmeyer *et al.* (1993) are closer to those occurring under natural conditions than those of Gubernick (1980, 1981b), and provide support for the former's suggestion that individual signatures of the offspring serve as a basis of social recognition. In that respect, the situation in the goat appears very similar to ones described in the following section on maternal selectivity in sheep.

Experimental studies on maternal selectivity in sheep

The ecological/life history variables of the domestic sheep are similar to those described in the previous section on the domestic goat. Like the doe, the pre-parturient ewe removes herself from the herd when giving birth. They also nurse only their own young and reject alien lambs. The mechanism of such selectivity is that of olfaction. Ewes are strongly attracted to birth fluids evidenced by their sniffing and licking of the ground where the waterbag has ruptured. Immediately after expulsion, the ewe licks her newborn which is covered with amniotic fluid and continues until the lamb is dry. This behaviour is associated with low-pitched bleats and pawing, which stimulates the activity of the neonate. As the lamb stands and searches for the udder, the mother responds by arching her back and making the teats accessible. Nursing behaviour is progressively organised into well-defined patterns where the lamb passes near the head of its mother to reach the udder, and the ewe sniffs the lamb's ano-genital area during suckling (Poindron, 1974). Normally, female sheep are strongly repelled by the smell of amniotic fluid, but at the moment of parturition they immediately become attracted to it. The duration of amniotic fluid attractiveness is a function of the mother's motivational state which is influenced by her contact with the lamb (Lévy, 1985, cited in Porter & Lévy, 1995). Ewes rendered anosmic by irrigating the nasal mucosa with zinc sulfate are neither repelled nor attracted to amniotic fluid.

Another method to test the assumption that attraction to amniotic fluid mediates maternal responsiveness involves the removal of it from the newborn's coat (Lévy & Poindron, 1987). Washing the neonate with soap and water before introducing it to its mother results in the following. Primiparous (i.e. first time) mothers show disruption of maternal responsiveness as reflected by little licking behaviour and much aggressive behaviour directed to the lamb. However, multiparous (i.e. experienced) mothers are able to compensate for the loss of olfactory information by relying on other cues associated with newborns, such as auditory and visual cues. Although the latency and duration of maternal licking are decreased, other aspects of maternal acceptance are not affected.

Once a selective bond is formed between the ewe and her lamb(s), two types of maternal discrimination of offspring may be observed – recognition at a distance, which permits the ewe to locate her young, and recognition at close quarters which is necessary before the young is allowed to suckle. The sensory basis of distant recognition of lambs is dependent upon vision and audition although to a lesser degree olfactory cues may also be salient (Alexander & Shillito, 1977). In contrast, the mediation of recognition at close quarters is mainly dependent upon olfaction. In general, the results of the studies of individual mother–infant recognition and care among ungulates demonstrate how this process is influenced by the ecology and life history of these species. A similar conclusion may be drawn from the results of studies done on rodents.

Maternal behaviour in rodents

A comparison between the different reproductive strategies of two species of rodents from the family Muridae – the Norway rat and the Egyptian spiny mouse – illustrates the principle observed in the ungulate studies. Norway rats are found throughout much of the world and live commensally with humans (Eisenberg, 1981). Their pups are altricial and incapable of leaving the burrow until two weeks old. These pups show selective response to olfactory cues emanating from lactating females, and there is a close synchrony of the time course in production of maternal pheromone and the age range over which pups are attracted to such cues. Mothers cease to emit the maternal pheromone about 27 days post-partum when their young are no longer responsive to such olfactory cues and when weaning is complete (Leon & Moltz, 1972).

In contrast to the Norway rat, the spiny Egyptian mouse inhabits rocky outcroppings in arid deserts, and gives birth to offspring whose sensory and motor systems are well developed. Unlike most small rodents, spiny mouse mothers do not construct a nest in which their pups are confined during the early post-partum period. At birth the eyes and ears are open and the precocial pups are capable of independent locomotion (Brunjes, 1988). The pups studied in the laboratory begin to approach home-cage bedding material when only one day old and eat solid food at four days. Similar to rat pups, spiny mouse young begin to move toward sources of maternal pheromone when they become capable of independent locomotion and are at risk of wandering away from the mother or home area. Attraction of the pups to maternal cues is a mechanism for ensuring necessary mother–infant unions. These olfactory cues act as 'beacons' that remain functional in rocky shelters. Similarly, rat pups are attracted to maternal odours in the vicinity of the nest. However, there are differences between the altricial rat and precocial spiny mouse with respect to age at which the pups orient to such chemical signals, as well as the timing of maximal production of maternal pheromone and developing motor capabilities.

Maternal selectivity is not as evident in rats and spiny mice as it is in ungulates. A classic experiment by Beach & Jaynes (1956) established that when rat mothers are introduced to alien pups they will accept them and nurse them. However, these mothers retrieve their own offspring more rapidly than alien pups. Acceptance of foster young, despite an ability to discriminate between them and one's own offspring, is consistent with the assumption that wild-living rats keep their newborn litters in individual nests isolated from other conspecifics. Porter & Lévy (1995) suggested that in this context, caring for any young in a nest is a viable maternal strategy because these young are likely to be the mother's own offspring. They also concede that that the ready acceptance of foster young may be an artifact of domestication of laboratory rearing conditions. Laboratory rats are fed a common diet which, in turn, results in less variability in odour cues among these mothers than those in the wild.

An experiment by Porter & Doane (1978) dealt with the responses of laboratory-housed spiny mice to alien conspecific neonates. When two females and their newborn litters are housed together in a large enclosure beginning shortly after parturition, these mothers nurse the alien young as frequently as their own. They also indiscriminantly retrieve their own and alien newborns in tests conducted when they were two days old. When the pups are eight days old the mothers interact preferentially with their offspring rather than the alien pups of the same age. However, they will still accept and nurse alien pups that are less than 24 hours old (Porter, Cavallaro & Moore, 1980). All the mothers in these experiments were maintained on the same laboratory diet. A different pattern of results was obtained in similar studies conducted with mothers whose diets were systematically varied. Porter & Doane (1978) found that their test subjects retrieved alien one-day-old pups whose mothers were fed a common diet more rapidly than pups whose mothers were fed a distinctly different diet. The olfactory signature of stimulus of pups may reflect the particular diet fed to their mothers. As with ungulates, dietary-dependent odours could be deposited directly onto the rodent pups as their mother licks them, as well as through the milk.

Maternal selectivity of rabbits

The European rabbit has spread, with human assistance, throughout the world. Wild rabbits inhabit warrens containing a system of burrows constructed by the colony members. Within a warren small subgroups establish individual territories that are protected against intrusion by others (Mykytowycz, 1985). Birth takes place in a separate burrow dug by the mother, and these chambers have only a single entrance, which the mother blocks with soil and marks with faeces and urine. Altricial rabbit pups are born following a 31-day gestation period and remained confined in the nest until approximately 13–18 days old. Unlike most mammalian mothers, female rabbits leave their pups alone in the nest and return only for a single period each day to feed them (Zarrow, Denenberg & Anderson, 1965). This unique form of intermittent maternal care may be an adaptation to heavy predatory pressure (Hudson & Distel, 1982). By blocking their young in an underground chamber and only visiting them for a short period daily, mothers reduce the risk of attracting animals that prey upon them.

Because the mother rabbit's daily visit to the nest is brief, the pups must rapidly locate her nipples and begin sucking, otherwise they will starve. The pups are highly dependent upon olfactory cues for nipple localisation and the emission of rabbit nipple-search pheromone is under the influence of reproductive hormones. Research on the odour of rabbit pups on maternal selectivity indicate striking results. Mykytowycz & Dudzinski (1972) observed the responses of captive does to three categories of pups introduced into their home pens. They were the female's own offspring, alien pups born in the same captive colony and wild-caught aliens. The mothers sniffed their own pups more often than either category of aliens. They rarely showed aggressive attacks to their own pups but attacked aliens by biting and pipping them with their claws. Over the course of the experiment, 21 per cent of the wild-caught aliens were killed through female aggression while their own pups were never killed. Odour cues for offspring recognition mediate the mother's differential behaviour. When their own pups were smeared with glandular secretions or urine from a strange lactating female, the females attacked them as if they were strangers (Mykytowycz, 1985).

SYNOPSIS OF PORTER AND LÉVY'S COMPARATIVE ANALYSIS

The analysis of the characteristics and the circumstances of maternal behaviour in goats, sheep, rats, spiny mice and rabbits reveal the intricate interaction of ecology, early experience, olfactory factors and learning as proximate variables that shape the pattern of maternal behaviour of these animals. The functions and ontogenetic patterning of olfactory signals exchanged between mothers and infants reflect complex interactions among these four groups of animals. Comparisons among them reveal the correlation between ecological/life history variables and particular parameters of mother–infant interaction. For example, precocial newborn lambs and spiny mice are not confined to a nest and are capable of wandering from the mother or home area. The attraction of these neonates to maternal cues ensures contact with their mother. With lambs visual cues allow these grazing animals to maintain social contact over relatively great distances when maternal odours are less salient.

As the opportunity for erroneous maternal investment in alien young increases, or when neonates put themselves in jeopardy by approaching alien females, enhanced individual discrimination would result. Lambs that indiscriminantly approach alien females would be delayed in becoming reunited with their own mothers and also risk physical attacks. However, such discriminative responsiveness would be of little adaptive value to newborn rabbits who are unable to leave the nest chamber and whose survival is entirely dependent upon the mother's daily visits. These pups are guided by stereotyped, rapid, inborn responses to odours emanating from the mother's milk and nipple region.

Because ungulates congregate in large groups, early offspring recognition enhances the mother's productive fitness by allowing her to invest her limited resources in her own young rather than wasting them on alien young. In a similar fashion, rejection of conspecific pups labelled with the odours of alien females is an adaptive strategy for mother rabbits because pups carrying such foreign labels would normally not be the mother's own offspring. However, the weak evidence for maternal specificity in laboratory studies of rats and spiny mice suggests that these mothers and their newborn live in isolation from conspecifics. An alternative explanation concerns the nature of standard laboratory housing conditions. Because all members of a breeding colony are fed the same diet, salient sources of odour variability necessary for individual discrimination may be eliminated.

PROXIMATE MECHANISMS MEDIATING MATERNAL BEHAVIOUR IN RATS

Much of the research on psychoendocrinology of maternal behaviour stemmed from the work of J. S. Rosenblatt and his associates at Rutgers University, New Jersey, USA. Most of their research was done on the rat and involved detailed descriptions of its maternal behaviour, the hormones that influence it as well as regions in the brain where they exert their effects. Prior to birth the pregnant rat builds a nest into which the young can be placed. When she gives birth to her first litter, she shows a complex pattern of behaviour made up of many different responses. She helps the expulsion of the young by pulling them out with her mouth, and also pulls out the amniotic sac. She eats the amniotic sac and attached umbilical cord and placenta and licks off the pup. She retrieves the pups to the nest site and once they are piled up together, she assumes a crouch posture and they attach to her teats and begin to suckle. Maternal behaviour persists, although with decreasing intensity, throughout the post-partum period until the young are weaned at 20–25 days.

In contrast to new mothers who are very responsive to newborns, even if they are foster pups (Fleming & Rosenblatt, 1974), non-mated virgin female rats show a different pattern of behaviour. Some virgins approach the pups, sniff them and then withdraw, and others become frantic after approach and cover the pups over with sawdust bedding. In extreme cases, these rats attack and cannibalise the pups. However, if a new litter of foster pups is placed in the female's cage each day and left for a 24-hour period, the female will, over time, begin to show behaviour very similar to that of the new mother (Rosenblatt, 1967). She will retrieve, lick and adopt a nursing posture over the young, although she does not lactate. This process known as pup sensitisation or stimulation involves two phases. During the initial phase the female remains at a distance from the pups and avoids them. The final phase occurs a few hours to a day post-partum, during which the female is in close proximity to the pups. During this phase the female becomes sensitised to the pups and shows the full repertoire of maternal behaviours.

Most of the experiments use latency in pup retrieval as the primary measure of the animal's maternal responsiveness. Thus post-partum rats who are very responsive to young pups do so without delay, whilst virgin animals have an average latency of about eight days. Some manipulations augment responsiveness in the virgin, thus shortening latency onset, and others prevent or decrease maternal responsiveness, thus increasing onset latency. The manipulation of greatest physiological import for this behaviour is the pattern of hormonal changes which occurs during the latter half of pregnancy. If ovariectomised virgin females are given the sequence of rising oestradiol and falling progesterone levels that simulates the hormonal changes of pre-partum females during this period, they will respond maternally within a day of exposure to the pups (Bridges, 1984).

Although maternal responsiveness is induced by hormonal factors, maintenance of this behaviour is dependent upon another variable, pup stimulation. This was demonstrated by Orpen *et al.* (1987) who implanted ovariectomised virgin rats with Silastic capsules containing oestrogen and progesterone in proportions that simulate the endocrine profile of pregnant rats 14 days prior to giving birth. Following these daily injections, the rats were either tested immediately or seven days later. Although the former group responded with short-latencies to the test pup, the latter group responded with long latencies that were similar to those exhibited by controls who were denied hormonal injections. These findings help us understand the classic finding (Rosenblatt & Lehrman, 1963) that non-treated rat mothers who show immediate responsiveness to their pups at birth, show a decline in responsiveness over the first post-partum week in the absence of daily contact or stimulation from their pups. Such results suggest that under normal conditions, hormonal factors are involved in the initiation or induction of maternal behaviour and that nonhormonal factors such as pup sensitisation arising from interactions between the mother and pups are responsible for the maintenance of this behaviour (Rosenblatt & Siegel, 1981). In general, once the mother rat expresses maternal behaviour as a result of hormones, she gains chemosensory and somatosensory experiences that sustain maternal responding beyond the early post-partum period of hormonal priming (Morgan, Fleming & Stern, 1992). Fleming, Korsmit & Deller (1994) found that experience with pup odours alone is not sufficient to sustain responsiveness. The female must also receive close physical contact that allows for both chemosensory and somatosensory stimulation.

There is evidence supporting the contention that the mechanism of the pup sensitisation effect is not hormonal. The finding that pup retrieval can be induced in castrated male rats confined in a cage with newborn pups for six to seven days (Rosenblatt, 1967) is consonant with this notion. Normally, males do not engage in parental care. The induction of pup sensitisation in males suggests that the basic capacity for maternal behaviour is present in both sexes. Although the neural network underlying maternal reactions may be prewired, the behavioural expression of this potential is latent. These reactions become manifest either because of the influence of the pre-partum sequence of hormonal changes on the relevant sites in the nervous system or because these sites are activated by a complex of stimulus cues emanating from the pups. Either of these factors reduce the threshold for the pup stimuli to elicit maternal reactions. The finding that male rats have the potential to exhibit maternal behaviour is intriguing. It may be related to the notion that 'female is the default program' for sexuality discussed in Chapter 2.

The results of blood transfusion studies also support the notion of a non-hormonal mechanism underlying the pup sensitisation effect. Although the latency of pup retrieval is reduced when blood from pre-parturient females is transfused into ovariectomised virgin female rats (Terkel & Rosenblatt, 1968, 1972), transfusions of blood from pup-sensitised rats does not reduce the latency of maternal behaviour in the recipient. These results along with those on the induction of pup retrieval in male rats clearly indicate that the pup sensitisation effect is *not* mediated by hormonal factors. It is just as easy also to induce pup retrieval in females whose pituitary gland had been removed as it is in intact females. A summary of the factors involved in rat maternal behaviour is presented in Fig. 3.1.

Fleming's model of hormonal influence on maternal behaviour

Fleming (1986) proposed that all females possess an underlying readiness to express maternal behaviour. Whether or not such behaviour is expressed depends upon whether or not there are competing responses which prevent or inhibit the appearance of these behaviours. Hormones act by removing the interfering influence of these tendencies in maternal behaviour. Virgin female rats actively avoid contact with the novel stimuli, particularly the odour of newborn pups. This active avoidance reflects a general difference in overall timidity between the virgin and the new mother. To test this hypothesis, Fleming compared virgin and post-partum animals on a variety of measures of timidity and found large differences in many of them. When tested in the open-field, post-partum females emerge more rapidly into the open, unfamiliar arena, ambulate more in general as well as enter the centre area more readily than do virgin females. Fleming believed that these differences in overall timidity reflect the influence of the parturitional hormones – oestradiol and progesterone. When the virgin females were implanted with the same Silastic regimen that had induced maternal behaviour in the experiments of Bridges (1984) and then tested in the open-field, their timidity was reduced to the level found in the post-partum animals.

The hormones also influence the female rats' olfactory reactions to cues from the pups. Whereas a new mother is attracted to the odour of pups, a virgin rat is repelled by it. As a test of this hypothesis, Fleming (1986) studied the preferences of these animals for the odour of nesting material taken from a lactating female and her pups. When the groups were given a Y-maze preference test with pup nest odour at one arm and virgin nest odour at the other, the post-partum rats prefered the former

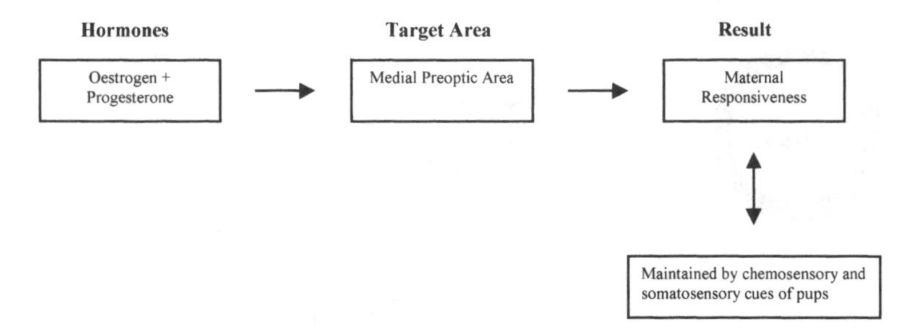

Fig. 3.1 Factors involved in the induction and maintenance of rat maternal behaviour.

odour while the virgins did not. When the ovariectomised virgins were injected with hormones that induce maternal behaviour, they showed a preference for maternal nest odours associated with pups. These results suggest that the post-partum females are willing to respond maternally to pups because they find their odour attractive. The importance of olfactory influence is revealed when it is impaired through removal of the olfactory bulb or damage to the cells of the nasal passage of the virgin rat following a zinc sulfate spray. In contrast to non-treated virgin rats, those made anosmic did not show any of the avoidance responses normally exhibited by the former.

Neural and hormonal control of maternal behaviour

The olfactory bulb is part of a region that forms the neural circuit regulating the expression of maternal behaviour. This circuit constitutes the primary projection route of the olfactory fibres and connects with parts of the mid-brain called the limbic system that are implicated in the regulation of emotional behaviour. A schematic representation of these components and the pathway is shown in Fig. 3.2, and a sketch of the location of these regions in the rat brain is presented in Fig. 3.3. The amygdala is a limbic structure that mediates affective/avoidance responses and evaluates stimulus salience. Lesions of this area reduce timidity in tests of neophobia (fear of novel stimuli) and open-field activity. The amygdala is connected to another neural structure called the bed nucleus of stria terminalis (BNST). These are primary projection sites of the olfactory system that mediate affective reactions to odours. Animals with lesions to these sites are not anosmic and can smell the pups but still respond maternally. This suggests that pup odours inhibit maternal reactions in intact animals.

The BNST is connected with the medial preoptic area (MPOA) of the

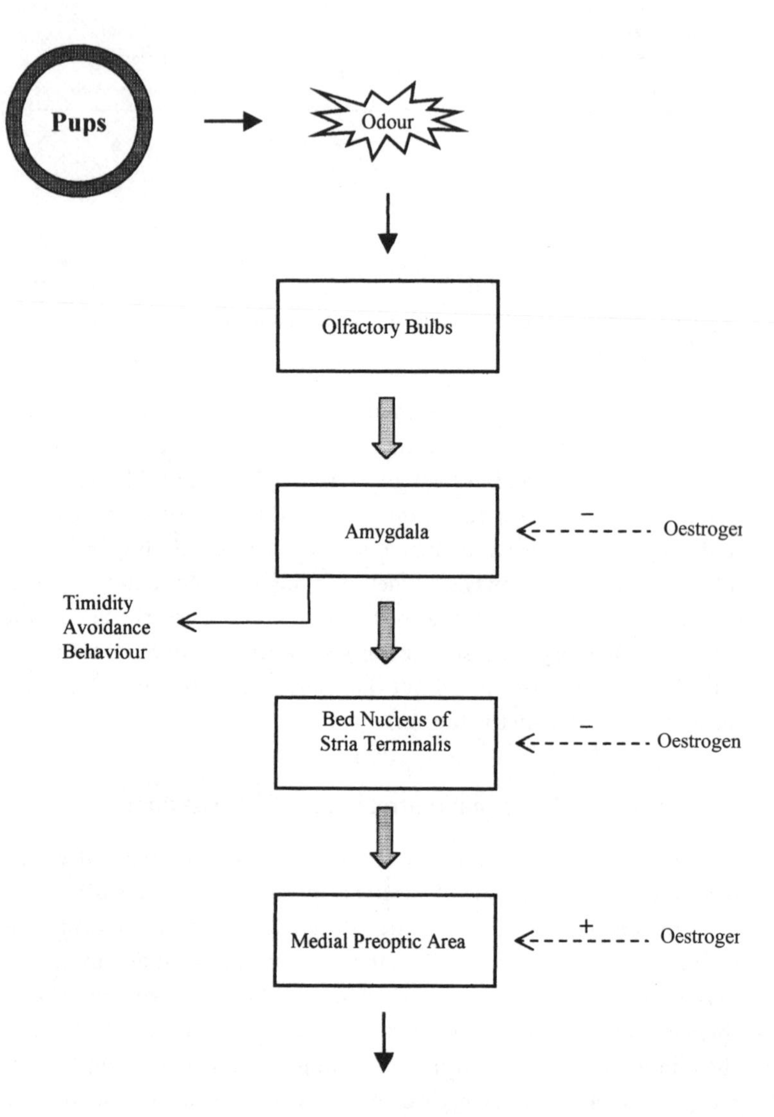

Fig. 3.2 Fleming's (1986) model indicating olfactory–limbic–preoptic pathways mediating rat maternal behaviour.

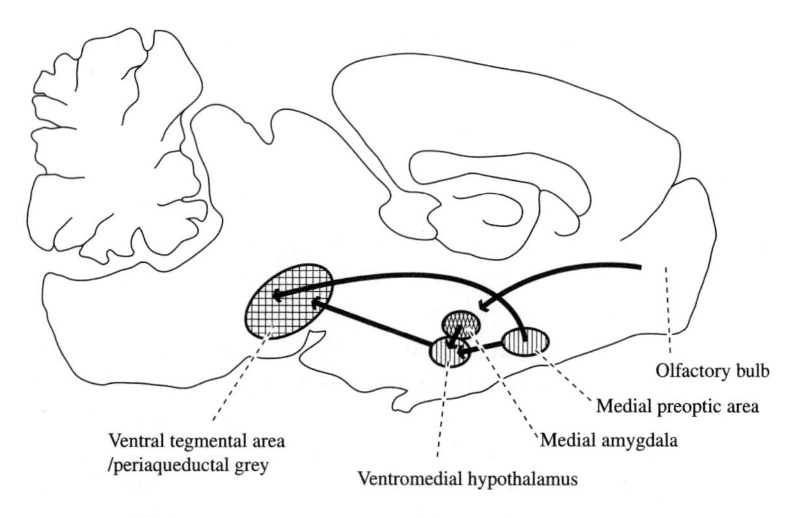

Fig. 3.3 Some of the brain regions implicated in Fleming's (1986) model. (Redrawn from Rolls, 1999, copyright 1999, by permission of Oxford University Press.)

hypothalamus, a region shown by Numan, Rosenblatt & Komisaruk (1977) and Numan (1996) to be essential for the expression of maternal behaviour. This behaviour is eliminated in post-partum rats following lesions of the MPOA or knife cuts and other lesions that transect its connection with other projecting fibre systems. However, rats with lesions in both the MPOA and the amygdala regions show no avoidance of pups and remain close to them, although they will not respond maternally. Upon the basis of her findings, Fleming (1986) suggested that the amygdala's primary function is to interpret olfactory input from the pups in terms of its novelty and salience and to make the appropriate affective response. When the external pup stimulus is novel and strange, the amygdala becomes activated and causes withdrawal and avoidance. With repeated exposure to the pup, the pups become less novel and more familiar, amygdala activity declines and the rats show less avoidance. If either the olfactory bulb or amygdala is lesioned, avoidance is reduced. In addition, the amygdala inhibits MPOA activity and when this inhibition is reduced, rats are able to express maternal behaviour.

The effects of hormones on maternal behaviour in post-partum animals are explained as follows. All the neural sites influencing maternal behaviour, preference for pup odours as well as timidity bind oestradiol. Bridges *et al.* (1990) and Numan *et al.* (1977) found that oestrogen implants in the MPOA facilitated maternal responding. The hormones in females during their preparturient period act not only to prime maternal behaviour and decrease the tendency to avoid them, but also to increase attraction to pups and their odours. Research done during the 1990s has also indicated the importance of prolactin in the induction of maternal behaviour in female rats (Bridges *et al.*, 1990; Bridges & Mann, 1994; Bridges *et al.*, 1997). During lactation, their endocrine system may be involved in the modulation or adjustment of the tone of the female's maternal responses. Prolactin is also involved in the inhibition of endogenous opioids, substances that inhibit active parenting in female rats (Hammer & Bridges, 1987). Another hormone that is released during parturition and nursing is oxytocin, and its effects on maternal behaviour were first reported by Pedersen & Prange (1979). This hormone is responsible for uterine contraction during parturition and milk ejection during nursing, and exogenous administration of it can induce short-latency maternal responding within minutes (Insel, Young & Wang, 1997). Insel (1990) noted that oxytocin interacts with both prolactin and endogenous opioids by increasing the secretion of the former.

There are also neurochemical or neurotransmitter changes accompanying parturition that affect hormonal release, and thus maternal behaviour. In Chapter 4, I discuss various neurotransmitter systems relevant to motivated behaviours, and in Fig. 4.5 a selected class of neurotransmitters is listed. The noradrenergic system shows enhanced activity at the time of parturition synchronously in widely separated neural regions such as the olfactory bulb, hypothalmus and substantia nigra (Keverne, 1988). This results in the synchronous enhancement of all the neural subsystems that are active in the context of maternal behaviour. Hence, maternal behaviour is potentiated as a consequence.

Fleming's (1986) model highlights the manner in which sensory stimuli, emotionality and motivated behaviour are interconnected. The proximate mechanism regulating maternal behaviour involves a neural system with excitatory and inhibitory functions, hormones and neurochemicals that excite or inhibit these neural components. Behavioural observations indicate that during their pregnant and immediate postpartum state, females are less fearful and more aggressive towards intruders than they would normally be. The adaptive function of this change in their motivational state is demonstrated by the willingness of the mothers to risk danger to themselves in order to protect their young. The proximate mechanism underlying this motivation involves behavioural dispositions that are activated as a result of physiological changes.

PARENTAL/MATERNAL BEHAVIOUR IN PRIMATES

Although the number of psychobiological studies on maternal behaviour in primate mammals is smaller than those concerning non-primates, the findings are somewhat congruent given the variability among primate species. Similar behavioural changes occur at parturition where the mothers are immediately attentive to their newborns. They are especially sensitive to auditory, tactile and visual stimuli from their infant(s). Such observations suggest that behavioural changes at parturition are hormonally induced. However, in contrast to results indicating that maternal responsiveness develops during pregnancy in rats exemplified by increases in nest-building and aggression toward conspecific adults (Rosenblatt, Mayer & Giordano, 1988), there is little evidence of changes in the maternal motivation of pregnant primates. Gibber (1986) found that pregnant primiparous rhesus monkeys do not show any increase in maternal behaviours such as picking up or remaining in contact with a 1- to 15-day-old infant presented to them. Rosenblatt (1991, p. 214) postulated that 'whatever effects hormones have emerge as specific behavioural changes during parturition itself and in the immediate post-partum period'.

We have reviewed the studies on how circulating levels of progesterone and oestradiol stimulate the onset of maternal behaviour in rodents and ungulates. Rosenblatt (1991, p. 207) suggested that the same pattern of hormonal changes at the end of pregnancy and the onset of maternal behaviour occurs among non-human primates as among non-primate mammals. The experiment by Pryce *et al.* (1988) on red-bellied tamarin monkeys provides indirect support for this suggestion. They found that the quality of maternal behaviour, measured in terms of post-partum infant survival, is associated with the level of urinary oestradiol during the last month of pregnancy. Mothers whose young survive for at least a week after parturition have significantly higher urinary levels of oestradiol at four to five weeks pre-partum than do mothers whose infants die during the first week. The latter mothers have lower levels of oestradiol which decline significantly during the first pre-partum week. The major consequence of this difference in hormonal pattern and maternal behaviour is survival rate of the infants. During a two–hour observation shortly after parturition, mothers with higher levels of the hormone lick, clean, carry and nurse their infants, whilst those with declining oestradiol levels push their infants off, and their infants die as a result of maternal neglect.

In general, data on non-primate and primate mammals indicate that the onset of maternal behaviour is hormonally based. The maintenance of these activities is regulated by different mechanisms. Rosenblatt, Siegel & Mayer (1979) proposed that the immediate post-partum period is one of transition in which the hormonal stimuli wane and new kinds of stimuli become the basis for long-term maternal responsiveness, namely those arising during interaction with the newborn. This is where the pup-sensitisation effect becomes salient. At later times, multiparous females may initiate maternal behaviour even when they are not exposed to the hormones which originally stimulated such behaviour in primiparous females.

Although primates and rodents are influenced by many common factors in the initiation and maintenance of maternal behaviour, there are striking differences between rats and primates in maternal attachment. In their comparison between the squirrel monkey and the rat, Levine *et al.* (1978) surmised that the rat mother does not manifest specific attachment to her offspring as does the squirrel monkey. This was demonstrated by measuring changes in corticosterone level, which reflects the state of arousal, following separation of the infant from its mother. The rat mother does not show any increase in plasma corticoid level (Smotherman *et al.*, 1977) in contrast to the elevation of this hormone in squirrel monkey mothers after a 30-minute separation from their infants (Coe *et al.*, 1978). These results are consistent with the behavioural observations by Levine & Wiener (1988) on the rhesus monkey. Removal of infant monkeys from their mothers causes distress to both parties. In addition, there is a weakening of attachment if they are separated for a substantial period of time. When they are reunited, such mothers show less maternal responsiveness than those who are allowed to maintain contact with their infants.

Although the rat mother does not manifest a specific attachment to her offspring, she responds differentially to distress stimuli emitted by the pups. However, when she is reunited with the pup, her pituitary–adrenal response remains high. This result contrasts with the data obtained with the squirrel monkeys showing amelioration of the response to disturbance when the mother and infant are reunited (Levine *et al.*, 1978). Such differences in the pituitary responses of these animals reflect different aspects of the mother–infant interactions in each species. The rat mother does not respond specifically to the absence of her own litter but does react if her pups are disturbed. In contrast, a specific attachment relationship develops between the squirrel monkey mother and her infant, and the arousal induced by separation is reduced by reunion with her object of attachment. Mendoza *et al.* (1980) suggested that the alteration of the mother's arousal by the infant may be an effective mechanism for inducing the attachment process in the mother.

One may wonder if there is any evidence for male attachment to infants and caregiving amongst non-human primates. In their investigation of social preferences and emotional bonds amongst the mother, father and infant titti monkeys, Mendoza & Mason (1986) and Hoffman *et al.* (1995) found that by the third month of life, the father is established as the infant's primary attachment figure, although the parents did not differ in their attraction to the infant. These monkeys are monogamous, so the results may not be typical of other primates. Yet Snowdon & Suomi's (1982) review of the literature on primates of many families and a variety of mating systems, indicate that male interest in infants is much greater than originally suspected. Similarly, Thierry & Anderson (1986) reviewed the literature on primate adoptions and found that males were the main adopters of post-weaning orphans. In an experimental study, Gibber & Goy (1985) presented juvenile male or female rhesus monkeys with young infants, and found few sex differences in responses to infants. However, when an infant is presented to a male and a female together, the female takes charge of the infant. Snowdon (1990, p. 249) has argued that 'the critical difference between polygynous or promiscuous and monogamous species is not the issue of paternal certainty, but that females in monogamous groups are more likely to relinquish some control of infant care to fathers and other group members. This occurs in monogamous species where females are more likely to require active participation by other group members in order to rear infants successfully'.

PARENTAL/MATERNAL BEHAVIOUR IN HUMANS

The data on changes in the pregnant mother's maternal responsiveness are mixed. Initial studies (Carek & Capelli, 1981; Robson & Kumar, 1980) indicated that women pregnant for the first time are no more responsive to a baby than non-pregnant women. Through the use of physiological measure of responsivness, Bleichfeld & Moely (1984) found that pregnant nulliparous women show a more variable heart rate response to the pain cry of an unfamiliar baby compared to non-pregnant women who showed no change in heart rate. In addition, Fleming, Steiner & Anderson (1987) focused on a psychological measure – maternal attitude – to determine whether individual differences in the women's scores are related to differences in hormonal levels. With the exception of cortisol levels, other measures of hormone or hormone ratio and of maternal attitude constructs were not associated with the various phases of pregnancy. At three to four days post-partum mothers who have higher plasma levels of cortisol spend more time being physically affectionate and talking to their infants than do mothers with lower levels of this hormone.

The situation is different during the parturitional stage. Several studies have suggested that there is a period encompassing the first hours after birth when the mother is particularly sensitive to contact, especially tactual, with her newborn. Carlsson *et al.* (1978) found that mothers with extended infant body contact show more affectionate behaviour towards their babies on post-partum days 2 and 4 than those who had less contact. Grossman, Thane & Grossman (1981) compared mothers who had early and extended contact with their babies for a few hours post-partum to those who had a brief view of the baby at birth and five routine feeding exposures during the rest of the day. The latter showed more affectionate behaviour towards their baby on post-partum days 2 and 4 but these differences disappeared by the 10th day.

The above studies suggest that mothers become attuned to their infant's cues soon after giving birth, probably as a result of hormonal factors, and that experiences resulting from interacting with their infant may contribute to their increasing attachment to their infant. Such results are congruent with those from the non-human animal studies reviewed earlier. They indicate that early and additional exposure to the newborn soon after giving birth is reflected in heightened maternal responsiveness during the early post-partum period. However, other studies on humans indicate that the benefits, if observed, are limited to a few measures out of many others recorded. For example, Fleming *et al.* (1987) reported that mothers who had a longer separation before their first extended period of nursing showed fewer affectionate approaches to the infants, but showed as many caretaking approaches when interacting with their three-day-old infants. In his review of the literature, Larsson (1994) concluded that most studies do not support the hypothesis that the physiological events of pregnancy and the perinatal period directly act to produce elevated maternal responsiveness post-partum. These events are neither necessary nor critical for maternal responsiveness to develop.

Since the publication of Larsson's (1994) review, Fleming's group found results that provide some encouragement to those favouring the original hypothesis. Fleming, Steiner & Corter (1997) found that prior maternal experience and post-partum cortisol level explain a significant proportion of variability in the mother's attraction to newborn infant odours. They conducted an experiment where new mothers were asked to complete a pleasantness scale to provide an attraction score to different odourants as well as complete scales that assess their attitudes towards infants, caretaking and related topics. The mothers gave salivary samples which were assayed for salivary concentrations of cortisol. Those with higher cortisol concentrations were more attracted to their own infant's body odour as well as being able to recognise that odour. However, the cortisol level was not related to attitudinal measures of maternal responsiveness, but mothers that had had more prior experience interacting with infants did exhibit both more attraction to infant odours and more positive maternal attitudes.

In his discussion on determinants of motherhood in human and non-human primates, Pryce (1995) proposed the concept of maternal reward which consists of four subsystems: the potential to be attracted to infants (AT); to be made anxious by them (AX); to have an aversion to them (AV); and to be afraid of their novelty (NO). Maternal reward potential is a composite neurobiological state which determines the reward associated with interaction with infants, and the behaviour that the mother demonstrates toward her infant. Potential AT increases the value of maternal reward and is demonstrated in human parents by slight autonomic changes and positive emotions elicited by their infants smiling (Fleming, Corter & Steiner, 1995). Simultaneous experience of sight, smell and touch of the infant releases the neuropeptides and cause an increase in its attractiveness (Keverne, 1995). The maternal anxiety caused by infant crying is a potent motivator of primate maternal care. Post-partum women become more aroused by infant crying than do non-pregnant women (Fleming *et al.*, 1995). The finding that human maternal behaviour is positively correlated with post-partum levels of cortisol in primiparous mothers, suggests that AX is a motivator of this behaviour. Individual differences in mothers' aversion to infant clinging and crying may reflect their AV potential. Halperin (1996) found that child abusers demonstrate greater physiological arousal and report more aversion and less sympathy to infant crying than do non-abusers. Maternal neophobia at parturition is normally low or absent in women with previous maternal experience (Pryce, 1995), and Fleming's model and research on the influence of NO on maternal behaviour of rats provide us with the mechanisms underlying this effect. In addition, personality variables such as as introversion/extraversion and neuroticism are factors in humans that may be linked with neophobic tendencies and maternal motivation.

Age-specific infant stimuli are processed by the mother, and the outcome of this processing will be a quantitative amount of maternal motivation, defined as the reward value that a biological mother assigns to her infant. At any stage of the mother–infant relationship, the higher the reward value of the infant to the mother, then the greater the amount of time and energy that the mother expends on her infant as care. Maternal motivation can be expressed as the net value of the mother–infant relationship (AT + AX – AV – NO). A net positive value for maternal motivation would result in maternal care and a net negative value would result in maternal neglect or abuse (Pryce, 1995).

Up to now, the discussion has been focused solely on maternal motivation and caregiving behaviour. In the previous section dealing with non-human primates, the evidence suggesting that males also participate in interactions with infants was discussed. Snowdon (1990) suggests that a similar mechanism (of economics) might be operating in human behaviour. A survey of parental care in many cultures indicates that human fathers are more likely to become involved with their children when mothers are more able to make an economic contribution to the family by working in areas other than child care. In societies where females' food gathering provides 60 per cent of the family's nutritional needs, male involvement in infant care is extensive (Draper, 1976; Hewlett, 1987). A comparison between male involvement in parental care among the 'developed' countries that differ in opportunities for female employment is instructive. Mothers living in Sweden (Hwang, 1987), Great Britain (Lewis, 1986) and North America (Leibowitz, 1978) have opportunities to work outside the home. In these countries, male involvement in infant care is greater than is the case with men living in Italy (New & Benigni, 1987) and Japan (Schwalb, Imaizumi & Nakazawa, 1987) where mothers have little opportunity to work. In industrialised societies, parents not only supply children with food and shelter until they can cope on their own, but also provide education and consumer goods for 4–12 years after a child would have been independent in a preindustrial society; thus, the economic demands of child-rearing are high, leading to mothers contributing to economic support in many societies. When this happens, both parents must share infant care or hire others to provide such care. Paternal care becomes more likely the greater is the economic value of the mother outside the home.

MECHANISMS OF ATTACHMENT

The concept of attachment refers to the emotional, cognitive and behavioural status of the infant relative to its primary caretaker (Pryce, 1995).

An attachment bond is characterised by a selective approach and interaction with selective individuals, as well as a display of affective distress when separated from these individuals during acute periods. Social attachment serves to maintain close proximity and elicit care from a primary caregiver, which in turn increases the probability of the young surviving to maturity and reproducing (Bowlby, 1989, 1991). Attachment to the young is initially induced by a maternal predisposition which arises from hormonal actions on a neural substrate. These hormonal effects also inhibit competing emotional tendencies such as fear of novel properties of the newborn and allows for the expression of the maternal disposition. This disposition is maintained by non-hormonal factors arising from infantile contact and interactions. Certain adaptive reactions may be pre-programmed in the offspring and these emerge during a sensitive period in birds. For example, young birds show a species-typical attraction or bias to markings and motion, a predetermined motor response for following and a tendency to a change of emotional state in response to moving objects within their perceptual bias. Normally, the mother is the source of these markings, sounds and motion, to which the young become imprinted or attached. Because the mother is normally present after hatching and her features fit the offspring's predisposition, she is the object of the relationship. Maintaining close proximity with the imprinted object becomes an important goal because the absence of it leads to distress. Staddon (1983) suggested that 'chicks are clearly "comforted" by proximity to the imprinted object'. They cease making 'distress calls' and begin to make 'comfort calls' during contact with the imprinting stimulus. A variety of behaviours are employed to maintain proximity with the object of attachment and reflect aspects of this new goal-directed system.

Although not constrained by a well-defined sensitive period, primates are also influenced by early experiences that influence attachment. The primates' situation is different to those of birds because of their longer period of immaturity and the more complex forms of interactions between mother and infant. Thus the period in which attachment develops is longer and variable among species. There is also a shift in salience of sensory systems (from thermal to olfactory to tactile to visual–auditory) as the infant matures. Although there are many theories of attachment, they share common premises (Kraemer, 1995). The infant is born with a plan for the development of brain function, and it requires a 'good enough' environment in which to express all of the capacities of its endowment. There is also variation in endowment, and the expression of it is limited to the exhibition of fixed action patterns in response to stimuli usually encountered in the neonatal environment. One effect of the attachment object on the infant is to regulate the infant's behaviour and physiology (Hofer, 1987). With persisting harmonious or predictable interaction with the caregiver, the behavioural and physiological regulatory mechanisms of the infant's brain organise and tune themselves in relation to the model presented by the caregiver. Their eventual synchrony and harmony of function depend upon the environment in which they develop and the characteristics of the caregiver. This organisation and tuning of brain function is an effect of attachment (Kraemer, 1992).

Birds and primates seem to undergo similar developmental changes where novel stimuli elicit approach during the sensitive period of imprinting. However, if they are not exposed to such stimuli during this period, they react with fear and aggression when exposed to such objects at a later stage. This analysis is relevant to the explanation of 'stranger anxiety' in humans. This phenomenon emerges when the child is six to eight months old and the development of it is not dependent upon any particular experience involving trauma (Sroufe, 1977). Instead, it is dependent on the prior attainment of a certain stage in cognitive development in which an infant is capable of internalised expectation. The infant expects an approaching person to be the mother (the first novel object in its life), and when the stranger's face is perceived, the resultant discrepancy causes distress.

Both fearful and attachment behaviour may be viewed as 'goal-directed' in the sense that they have a predictable outcome. As described previously, a predictable goal of attachment behaviour is gaining and/or maintaining proximity or communication with an attachment figure (Ainsworth *et al.*, 1978; Bowlby, 1969). Data on human infants are consistent with those on imprinted birds in that the baby's heart rate accelerates upon separation from its mother and decreases once proximity is regained (Sroufe & Waters, 1977). The outcome of fearful behaviour is to decrease or avoid proximity to and/or interaction with the feared object. Sroufe and Waters (1977) found gaze aversion to a stranger occurs when heart rate acceleration was near its peak and that the baby looked at the stranger again when the rate had decreased. Protection from harm may be both a function of fear and attachment (Bowlby, 1991; Sroufe & Waters, 1977). Fear of the unfamiliar or of being left alone would have been essential for survival in the environment in which humans evolved (Stevenson-Hinde, 1991). Children's propensity to fear certain situations such as the dark or water or snakes and to maintain proximity with their mothers may have been guided by natural selection because it serves the function of protection from harm. Neurobiological models such as those of Hofer (1987) and Kraemer's (1992) explain these functions in terms of the state of behavioural homeostasis attained as a result of removal from these situations and proximity with the attachment object.

CHANGING PATTERNS OF PARENTAL INVESTMENT

Parents make decisions about how much to invest in their offspring at any given time. There comes a time during the offspring's development at which the infant seeks more care than the mother is prepared to give. The situation where the parents' interests do not coincide with those of their offspring led Trivers (1974) to propose the concept of 'parent–offspring conflict'. His reasoning was as follows. A mother both derives benefits and incurs costs by caring for an infant. She increases its chance of surviving and growing, thus benefits by the the increased number of progeny she leaves. Raising infants is demanding and very time-consuming, so by committing resources to rearing a particular infant or family, the mother is reducing her chance of successfully rearing infants in the future. As the infant grows bigger, the cost will increase, especially because the baby will demand more food. Moreover, the cost of not looking after the baby will decrease because the infant will become more self-sufficient. Thus, there comes a time when it is to the mother's advantage to stop caring for her infant and to start a new family.

Infants and mothers view the costs of providing maternal care from different perspectives. By demanding care from its mother, the infant reduces the chance of her successfully acquiring future brothers and sisters. To the infant, its own survival is more important than the survival of its siblings. Trivers's (1974) analysis may be illustrated by an examination of specific events occurring during weaning of animals such as rats, cats and dogs. As long as the mother is lactating, she is expending energy on her current litter. If her litter could be independent of her, then continuing to feed them would reduce her lifetime reproductive success. However, she is able to produce another litter only when her oestrous cycle returns, and this cannot happen while she is still feeding them. Rothchild (1962) reported that suckling by the young inhibits the pituitary release of gonadotrophins that stimulate oestrogen release, and 'the human female shows the same general relationship between the amount of daily breast-feeding and the length of post-partum amenorrhea'. Similarly, Anderson (1983) reviewed data indicating post-partum anovulation due to breast-feeding, and argued that the four-year birth interval of the !Kung hunter–gatherers of Botswana is entirely due to the contraceptive effect of lactation. Hence, at some stage the mother will curtail feeding of her current litter in order to be able to invest in the next set of offspring. This results in avoidance and outright rebuffing of the young which provides the basis of parent–offspring conflict. Rosenblatt (1990, p. 52) cautions us not to regard *parent–offspring conflict* as an explanatory term with reference to the process of weaning in the rat and other animals. The conflict referred to in Trivers's theory is *not* the behaviour often seen in litters in which the young try to nurse from the mother while she avoids and evades them (Rosenblatt, 1965). This is a behavioural conflict, and is different from the *conflict of interests* between the mother and her offspring over efforts to improve their individual fitness.

The exact time of weaning is affected by the age and health of the parent or by the availability of food or by signs that the offspring is ready for independence, such as its size or behaviour or the pitch of its cries. These cues are used by the parent to determine how much care to give its young. They are also cues an offspring can manipulate in order to cause the parent to give more care than it wants to. In humans, as soon as a system of parental attentiveness to signals from its offspring evolves, 'the offspring can begin to employ it out of context. The offspring can cry not only when it is famished but also when it merely wants more food than the parent is selected to give. Likewise, it can begin to withhold its smile until it has gotten its way' (Trivers, 1974). Although selection will result in the parental ability to discriminate between the two uses of the signals, it also will result in the development of subtler mimicry and deception in the child. Trivers suggests that at each and every phase of parental care, parents should be selected to make sure that the costs of care are as low as they can be in relation to the benefits. Although researchers do not know the mechanisms or 'physiological calculators' by which animals and humans arrive at these decisions, one thing is clear – only those animals or humans that make the right calculations survive and leave progeny behind.

Parent–offspring conflict among humans extends well beyond the weaning period, and is manifested in their differences of opinion concerning appropriate behaviour. In modern affluent western societies, parents and adolescent children often clash over parental prescription for well-being, and their disagreement concerns issues such study habits, smoking, gambling, premarital and unprotected sex, dangerous driving and the like. In general, parents favour their children engaging in useful activities such as studying, tidying up their rooms, and in the case of rural families, tending the animals, weeding the garden and the like. In contrast, children favour activities that involve 'frivolous' expenditure of energy such as playing, and perceive behaviours favoured by their parents as being dull, unpleasant and boring. Parents are concerned with the long-term consequences of their children's social behaviour which, in turn, have implications for their fitness maximisation. The children, in contrast, are driven by the 'pleasure-seeking' activities, and seem to be governed by recreational events providing immediate gratification such as alcohol and drug usage. Parents attempt to remind their children of the long-term effects of their activities, often with minimal success. Such change in attitude and behaviour seems to come about mainly through maturation. In the late 1990s research suggests an explanation involving proximate mechanisms responsible for the tendency of children to engage in behaviours that are the source of anguish for most parents and constitute the basis of much parent–offspring conflict.

Through the use of functional magnetic resonance imaging techniques, Yurgelun-Todd and colleagues completed a series of studies which suggested a possible physiological basis for the emotional turbulence of adolescence and the gulf of misunderstanding that sometimes separates the generations (D. A. Yurgelun-Todd, personal communication). Baird *et al.* (1999) compared young subjects between the ages of 9 and 17 years with those between the ages of 20 and 40 years on their handling of mental tasks involving emotion and language. We have already encountered analysis and discussion of the role of the amygdala in fear in Fleming's (1986) model of maternal behaviour in rats. In addition to its role in mediating fear-based responses, the amygdala is also important in the formation of emotional memories, and plays a role in forming intuitive social judgements amongst humans. By studying how each subject responded to a series of pictures containing faces expressing different emotions, the researchers found that when younger people process emotion, the level of activity in the amygdala is higher than the activity level found in adult subjects. Not only did the adolescents overreact in terms of involuntary mental activity, but they also could not correctly identify the emotions in the pictures they were shown. In another study, children and adolescents were compared to adults on a test of language function (D. A. Yurgelun-Todd, personal communication). When activation of the frontal and temporal lobes was examined, greater activation was found in the frontal lobes in adult subjects whereas greater temporal lobe activation was found in the adolescent group. Yurgelun-Todd (personal communication) examined the combined data sets and found that in childhood and adolescence there was an observed reduction in frontal lobe activation for both emotional and cognitive processing demands. The reduction in frontal lobe activation and increased temporal lobe activation may be associated with the frontal cortex, an area which has a tempering effect on judgements. These findings suggest that as adolescents mature into adulthood, they are more able to temper their emotional reactions with reasoned responses. The researchers suggested that such results may explain why adolescents often produce incongruous responses to emotional stimuli. Thus, children may not be understanding what their parents are telling them and may not appreciate the consequences of the actions. This leads to the inevitable parent–offspring conflict.

SUMMARY

Parental/maternal motivation involves the tendency to respond to infants in such a manner that promotes their well-being. Activities arising from this motivation ensures the survival of offspring and thereby enhances the parents' reproductive success. Amongst mammals, females bear the brunt of parental caregiving, partly because of the nature of the nursing relationship, and also because of confidence of parentage. The certainty of paternity is lower in males. However, in many monogamous species of primates, biparental care occurs when successful rearing of offspring is dependent upon the work of both parents. There is diversity in the extent of attachment between parents and offspring amongst mammalian species. Much is dependent upon the life history and ecology of the species.

The proximate mechanism regulating maternal behaviour involves a nervous system with excitatory and inhibitory functions, and hormones as well as neurochemicals that excite or inhibit these neural components. The onset of maternal behaviour is hormonally based, but the maintenance of these activities is regulated by a different mechanism, namely stimulus exposure. At a certain point in time, there is a conflict of interests between the mother and her offspring over efforts to improve their individual fitness. For example, the exact time of weaning is influenced by factors such as the age and health of the mother or by the availability of food or by signs that the offspring is ready for independence. At each phase of parental care, the mother behaves in such a manner that the costs of care are low in relation to their benefits. This may generate disagreement from the offspring as it matures.

4

Feeding activities

Feeding is the means by which an organism acquires the materials for building, maintaining and moving the vehicle that carries the next generation. Since nutrition is the main requirement of all living systems, feeding preceded by food-seeking behaviour, is a necessity of life. Organisms must feed to live and they also must work to feed. Evolution has played a role in influencing the eating and drinking behaviour of all species. It has selected for the mechanisms that motivate the organism's ingestion of nutrients and its selection. There are two fundamental aspects of feeding: energy balance and diet selection. Energy balance deals with how much animals eat in relation to their energy expenditures, whilst diet selection deals with mechanisms that allow omnivores to choose the appropriate nutrients. This chapter will deal with the former and Chapter 5 will be concerned with the latter issue.

Organisms regulate their nutritional intake according to short- and long-term energy needs. The mechanism responsible for this regulation, which involves the maintenance of a constant, optimal, internal environment, is called homeostasis. This concept was introduced by Claude Bernard (1879) in his discussion of '*le milieu intérieur*' and the necessity of the organism keeping its internal environment at a constant, optimal level. (For a detailed discussion of the key components of homeostatic mechanisms and their character of operation, see Schulze (1995).) An example that illustrates the concept of homeostasis is that of the thermostat. After it is set for a certain temperature, it reacts to deviations from that temperature (the set point) by changing the environment so that such deviations are eliminated. If the temperature is too high, the heat source is turned off, and if the temperature is too low, the heat source is turned on. The process by which deviations from a set point result in actions designed to reinstate the set point is called negative feedback. As the deviations are reduced, so are the reactions to

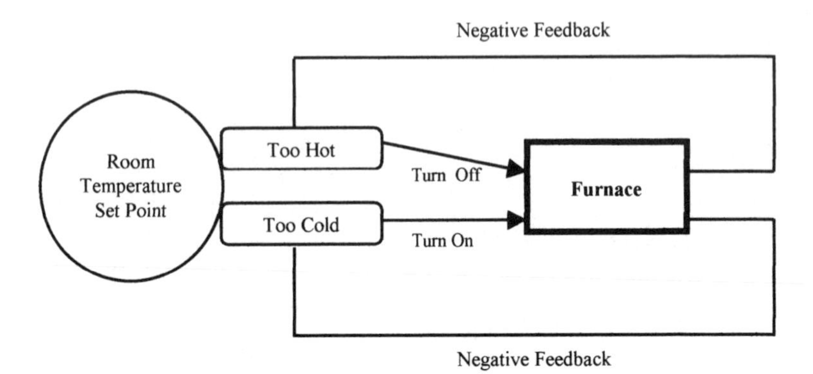

Fig. 4.1 Thermostatic control system involving negative feedback.

them, and a stable state is reached eventually and maintained. An illustration of this control system is shown in Fig. 4.1.

The thermostat illustration points out three fundamental components of homeostatic regulation. There is a parameter that is critical to the system and which can be regulated, a perturbation- or error-detecting mechanism and an effector mechanism that, when activated, corrects errors. For example, eating can be considered an effector response in the homeostatic control of plasma glucose (blood sugar). Woods, Taborsky & Porte (1986) found that this substance is normally maintained within a narrow range, and that a well co-ordinated series of physiological and behavioural responses are enlisted to maintain plasma glucose within this range through a negative feedback system. Effector control mechanisms can raise or lower plasma glucose, but under normal conditions, there is a balance between these opposing actions and little variation in the level of the critical parameter. As challenges arise, the control system can be rapidly recruited to counter changes in the critical parameter, and the system responds and restores the variable to its homeostatically defended level. However, such reliance on reactive measures is inefficient and can be avoided through learning. If through experience an animal could learn to recognise situations in which plasma glucose is likely to change, it could make an a priori adjustment in the ongoing homeostatic control over glucose and successfully minimise any changes that have occurred (Dworkin, 1993). This kind of control is called *feedforward*, because the animal is able to predict when homeostatic perturbations will occur and can initiate anticipatory reponses to deal with them. This type of regulation occurs efficiently in the absence of significant errors and the feedback arm of the system rarely needs to be activated (Ramsay *et al.*, 1996). Although there are differences in the extent to which contemporary theories of feeding emphasise these mechanisms, most consider the role of both feedback and feedforward factors.

PERIPHERAL CUES AND THE DIGESTIVE SYSTEM

The digestive process consists of food entering the mouth, continuing in the stomach and its nutrients being absorbed in the intestines. The digestive system is a logical place to begin the analysis of the amount and timing of food consumption. Since the stomach is one of the most obvious organs of food storage, earlier researchers thought that peripheral cues emanating from that region can predict the initiation and termination of feeding. Contractions of the stomach would predispose the organism to initiate feeding, and distension of it would result in the cessation of further intake. Early data, using a single subject design, suggested that stomach contractions were closely associated with the subject's own reports of hunger (Cannon, 1929). When the subject did not report the experience of hunger, there were no contractions. However, later studies using improved methods of measurement, have indicated that the relationship between reports of hunger and the appearance of stomach contractions to be weak (Stunkard & Fox, 1971).

The stomach was also thought to be involved in the termination of feeding through its distension. Arguments for this notion were based on the following findings. There are stretch receptors in the stomach wall that have inhibitory effects on feeding. Activating these receptors by inflating a small balloon within the stomach or constricting the emptying of the latter both suppressed feeding. Deutsch (1985) reviewed data indicating that signals of stomach distension are sent to the brain via the vagus nerve. Feeding suppression is more complete when nutrient receptors in the stomach lining are activated. However, when nutrients enter the stomach, chemical signals are also sent along the splanchnic nerves to the brain (Deutsch & Ahn, 1986).

The intestines are the next section of the system to work on the consumed food. They release a hormone, cholecystokinin (CCK), that suppresses appetite (McHugh & Moran, 1986). Intraventricular injections of CCK greatly reduce the amount of food consumed by rats (Gibbs, Young & Smith, 1973) and monkeys (Gibbs, Falasco & McHugh, 1976). Similarly, Zhang, Bula & Stellar (1986) also found that intraventricular injection of CCK depressed the intake of food by rats as well as its instrumental behaviour. In addition, the liver is involved in the regulation of feeding by monitoring the nutrients in the blood through its connection with the intestine. Large quantities of glycogen are stored in the liver and drawn upon to fuel the body. The liver contains receptors for several types of nutrients and communicates with the brain via the vagus nerve. These receptors increase their rate of firing when there is a decline in the nutrients monitored in the liver's blood supply. Friedman & Stricker (1976) proposed that hunger signals are sent to the brain when fuel from the intestine and fatty tissue 'is inadequate for the maintenance of body functions without significant hepatic contributions'. Blocking conduction in the vagus nerve reduces feeding in hungry animals.

The findings presented above indicate that the body uses various sources of information for determining how much should be eaten and how much has been eaten. The balance between the amount of energy entering the body and the amount of energy expended is regulated through homeostatic mechanisms. The regulatory system receives information from many parts of the digestive system which acts as signals about the energy state of the body. This information must then be integrated within some central control system that will also regulate the response. Short-term regulation of energy intake that influences the inter-meal interval and meal size may be controlled by factors other than those concerned with long-term regulation of energy reserves. The latter involve mechanisms that maintain the stability of body weight.

BRAIN MECHANISMS INTEGRATING FEEDING

Various brain regions collect information about the organism's energy state using sensory pathways. Based on such information, these regions could initiate or terminate feeding through motor nerve pathways. Most of the early research, done in the 1960s, on brain mechanisms of feeding focused on two regions of the hypothalamus – the ventromedial hypothalamus (VMH) and the lateral hypothalamus (LH). Although research in the 1990s has indicated the influence of interconnected regions, which we will examine later, the logic as well as the results of the approach used in these classic studies is worthwhile reviewing. Such studies provide us with a background for understanding alternative explanations and current developments in the field. Some of these developments focus on cerebral events and the role of learning in their influence on feeding, an approach which sets the stage for the *cephalic phase hypothesis* of obesity. We will also examine other recent concepts which explain the role of the hypothalamus from a broader more complex theoretical perspective.

Early studies of the VMH suggest that it was the site of appetite suppression (Stellar, 1954). This notion was derived from experiments in which the researchers inserted small electrodes into the brains of anaesthetised rats. When the tip of the electrode resided at the VMH, a direct current was passed through the electrode, thereby destroying the cells that lie around and below the tip. After the anaesthetic drug wore off, the rats woke up and encountered food pellets in their cages. The experimenter monitored the feeding patterns of the rats over time and compared them with the those of non-lesioned rats. Normally, rats eat about 12 small (1–2 g) meals, usually eight during the night and four during the day within a 24-hour period. In contrast, the VMH-lesioned rats ate more than 12 meals that were larger than the usual 1–2 g. This pattern of overeating is termed hyperphagia. Within a month or so after the operation, the VMH-lesioned rats may have more than doubled their body weights. Given these disruptive effects of VMH lesions on feeding, it is assumed that regulation of food intake is one of the direct or indirect functions of this neural region.

Effects of VMH lesions

More detailed analyses of the feeding patterns of VMH-lesioned rats indicate the following. During the immediate post-operative period, rats display an increase in general activity that lasts for several hours and then changes to chronic hypoactivity. Balagura & Devenport (1970) distinguished three periods: an acute dynamic period; a chronic dynamic period; and the static period. From the time the VMH-lesioned rat begins recovering from the anaesthesia, a single, extended bout of feeding occurs for about 12 hours. This persistent feeding gradually tapers into discrete meals. The experimenters hypothesised that the behaviour of VMH-lesioned rats is an exaggeration of the behaviour observed in non-lesioned rats that have been deprived of food for about 36–52 hours. When the latter is given access to food, they also exhibit a prolonged bout of feeding, but resume normal feeding after this binge. However, the VMH-lesioned rats are more ravenous because they show a more sustained period of prolonged feeding (the acute dynamic period).

During the chronic dynamic period, the general activity of the rats is greatly reduced, they are hyper-reactive to external stimulation and eat discrete meals more numerous than their normal 12. The interval between meals is understandably decreased as a result of the increase in the number of meals. After a period of time, the amount of which is a function of the size of the VMH lesion, the rats return to a pattern of eating 12 meals daily. Although their feeding pattern is similar to the one which they showed during the pre-operative period, they maintain the body weight level induced by their overeating during the preceding periods.

The fact that the rat has entered the static period does not mean that it has recovered normal food and body regulation. If it is deprived of food until its weight is reduced to the pre-operative level and then allowed free access to food, it will renew its voracious feeding until its weight reaches the predeprivation level (Hoebel & Teitelbaum, 1966). In essence, the rat seems to go through a second dynamic phase and recovers its obese weight. However, if it is force fed until it reaches a higher body weight level and then given *ad lib.* access to food, it will decrease feeding until it has attained the weight level it achieved prior to being force fed. If a non-lesioned rat is overfed to obesity and food is made freely available, it will voluntarily stop feeding until its body weight attains normality. With a VMH-lesioned rat, a constant body weight is attained, but at a higher than normal level.

Subsequent research has indicated that lesioned female rats show not one but two abnormal stages of weight gain (Hallonquist & Brandes, 1981b). During the first 10 post-lesion weeks the rats showed a gradual increase in weight followed by a phase in which they manifested continuous fattening. Following the 11th post-operative week, the VMH rats maintained a steady increase in body weight at a mean rate more than double that of the non-operated control group. Manipulation of the body weight of individual rats during the second phase produced results that suggest that the lesion induces a gradual elevation of set point with time, which the authors named 'a climbing set point'. Another study by Hallonquist & Brandes (1984) replicated the two phases of weight gain. When the VMH rats were food-restricted between the 20th and 26th weeks post-lesion, their rate of weight gain following release from food restriction was greater than that shown during the 11th week post-lesion. The authors suggest that VMH lesions induce a gradual climbing of the set-point for body weight that occurs independently of actual food intake. Mrosovsky (1990) used the term *rheostasis* to describe the condition or state involving regulation around shifting set points.

In addition to its effect on food intake, several changes in taste preferences are produced by the VMH lesion. The rat becomes hyperreactive to flavours. Although food intake is maintained despite adulteration of the diet with unpalatable substances during the dynamic periods, it is a different matter during the static period. The rats become finicky and reject diets containing a bitter-tasting substance such as quinine (Ferguson & Keesey, 1975). If kept on such an adulterated diet, the rats do not show hyperphagia nor obesity. If bitter diets are introduced after the rats become obese, the VMH-lesioned rats virtually stop feeding until their body weights approximate their pre-operative levels (Franklin & Herberg, 1974). Even changes in the texture of the food (e.g. powdered vs. pelleted food) affect their intake.

Not only is the rat hypersensitive to the negative aspects of the diet, but it is also hypersensitive to the positive aspects, and will overeat when offered highly palatable food (Teitelbaum, 1955). Similarly, Corbit & Stellar (1964) found that VMH lesions resulted in greater consumption of sweet-tasting foods and liquids as well as decreased intake of unpalatable foods and drinks. One interpretation of these changes is that the rat is less responsive than before to its internal state and more responsive to external influences such as taste. This notion was one that influenced Schachter's (1971) externality theory of human obesity.

Further support for the notion that the VMH-lesioned rat shows an exaggerated reactivity to sensory stimuli arise from experiments indicating that it is more sensitive to painful electric shocks than a control animal (Turner, Sechzer & Liebelt, 1967; Vilberg & Beatty, 1975). These lesions also facilitate the rat's performance on shock-motivated active avoidance behaviour (Grossman, 1966, 1972). Many experimenters have found that the lesioned rat shows greater resistance to capture than the control and is more likely to attack when touched. In an attempt to sort out the effects of VMH lesions on hyper-reactivity and obesity, Hallonquist & Brandes (1981a) manipulated the housing conditions of the lesioned rats. Although rats housed in single cages were more hyper-reactive to the experimenter than those housed in group cages, they did not differ in lesion-induced obesity. Such results suggest that VMH-induced obesity and its effects on affective reactions are not interdependent.

THE CEPHALIC PHASE HYPOTHESIS

One of the most provocative explanations of the VMH syndrome discussed in the preceding section suggests that lesions of the VMH affect palatability (Powley, 1977). From this perspective, good-tasting food tastes even better, and disliked food tastes even worse as a result of the lesions. The amount of food consumed is thereby changed as a result of the change in palatability. Because VMH lesions affect certain metabolic functions as well as feeding, Powley proposed that the latter is a consequence of the former. These lesions cause an exaggerated autonomic and endocrine response to food, responses that he calls *cephalic reflexes of digestion*. These reflexes, such as an increase in saliva or insulin release as a result of the taste or smell of food, lead in VMH-lesioned rats to overeating and other symptoms. These stimulus-response sequences are related to the ingestion and digestion of food and occur immediately upon the animal's direct contact with food. According to the cephalic phase hypothesis, destruction of the VMH causes an exaggeration of these reflexes, which in turn disrupt the energy balance of the organism.

The cephalic phase hypothesis differs from traditional explanations in its analysis of hypothalamic function. VMH-lesioned rats become obese not because they overeat, but because of metabolic activity induced by the lesion. The VMH regulates metabolism, and feeding is thereby affected. Lesions therefore cause the rat to become overresponsive to the taste (and other stimuli) associated with food and the rat, in effect, is locked into a positive feedback loop and hyperphagia is the result. The primary effect of VMH lesions is its effects on insulin release from the pancreas. This substance facilitates the transport of glucose into cells as well as converting glucose into a stored form of energy – fat. This process is called lipogenesis. Because the calories of VMH-lesioned rats are converted to fat at such a high rate, the rat must keep eating to ensure that it has enough calories in its blood to meet immediate energy requirements.

This interpretation of the VMH syndrome is supported by the following findings. Insulin levels in the blood are elevated after VMH lesions (Hustvedt & Lovo, 1972). Rats with VMH lesions convert more of their food to stored fat, even when they eat the same amount of food (Han, 1967). This food is absorbed more rapidly from the gut than in non-operated controls (Duggan & Booth, 1986). VMH-lesioned rats exhibit increased magnitude of gastric acid secretion (Weingarten & Powley 1980). The hypersecretion did not require hyperphagia or weight gain, but its magnitude correlated with subsequent body weight gain. Some physiological and behavioural effects of VHM lesions are presented in Fig. 4.2.

The vagus nerve, especially its efferent branch, is a major pathway that transmits signals from the brain to the pancreas. The elimination of the cephalic reflexes in VMH-lesioned rats should also eliminate the rat's hyperphagia. This can be done by either cutting or pharmacologically blocking the vagus nerve. Several experiments have confirmed that such manipulations decrease VMH-lesioned rats' eating and obesity (Cox & Powley, 1981; Eng, Gold & Sawchenko, 1978). These experiments involved sectioning the vagus nerve just below the diaphragm, and unlike

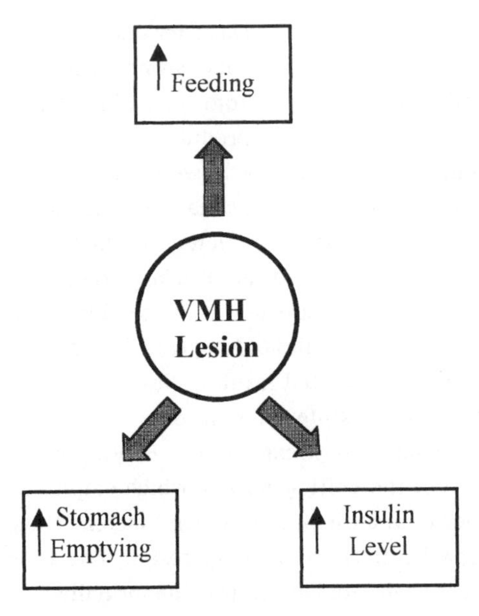

Fig. 4.2 The physiological and behavioural effects of ventromedial hypothalamus (VMH) lesion.

the control VMH-lesioned rats, the vagotomised rat regulates its body weight at normal or slightly below normal levels (Snowdon & Epstein, 1970; Snowdon & Wampler, 1979). The fact that VMH-lesioned rats did not show hyperphagia following vagotomy indicates that this neural region does not directly control feeding behaviour but may do so through its interaction with pancreatic activity.

The secretion of insulin from the pancreas is directly influenced by the dorsal motor nucleus of the vagus (DMNV). Although the glucoreceptors in the VMH and LH may not send axons directly to the DMNV, neurons of the nearby paraventricular nuclei (PVN) do (Gold, Jones & Sawchenko, 1977; Kirchgessner & Sclafani, 1988). Glucoreceptors in the tongue and liver send information to the PVN (Novin *et al.*, 1985) and possibly, glucoreceptors within the VMH and LH do likewise. This possibility could be tested by applying glucose to the LH or VMH and examining the effects on PVN activity.

Although the cephalic hypothesis provides an intellectually satisfying explanation of VMH hyperphagia, the results of some experiments reveal that hypothalamic obesity is *not* always prevented by prior vagal transections (King & Frohman, 1982). In their review of the literature, King and Frohman concluded that vagally-mediated hyperinsulinaemia can account for no more than 40 per cent of the weight gain observed in animals with VMH lesions fed *ad lib*. They also suggested that the obesity resulting from some VMH knife cuts may not be mediated by hyperinsulinaemia. Similar findings arose from Sclafani's (1981) experiments, revealing that VMH cuts do not produce hyperinsulinaemia when the rats are prevented from overeating. However, insulin levels are elevated when the rats were allowed to overeat. He also found that although subdiaphragmatic vagotomy completely blocks VMH hyperphagia and obesity if the rats are fed chow, it does not prevent overeating and rapid weight gain if they are fed an assortment of highly palatable food. The upshot of the studies of King & Frohman (1982) and Sclafani (1981) is that vagally mediated insulin release may not be an essential component to the VMH 'knife cut' syndrome.

In contrast to the cephalic phase explanation, Booth (1994, p. 171) proposed an explanation of the VMH syndrome which emphasises visceral mechanisms rather than anticipatory cerebral factors. Duggan & Booth (1986) and Toates & Booth (1974) obtained results indicating that VMH rats overeat because their stomach empties abnormally fast after the region has been destroyed. The VMH normally transmits inhibition of stomach-emptying through the autonomic nervous system. The speeding of absorption by uninhibited gastric emptying provokes extra insulin secretion to deal with the absorbed nutrients. This is how the extra food gets deposited as fat. After lesion of the VMH, insulin secretion is chronically increased within a few meals in response to the faster absorption and/or from direct loss of inhibition. Hence the VMH rat becomes obese even if it is not eating any more than a control rat. Intriguing though Booth's explanation may be, it does not account for some of the problems alluded to in the reviews by Sclafani (1981) and King & Frohman (1982).

Effects of LH lesions

Although one may question whether VMH lesions produce hyperphagia because this area directly affects feeding or cephalic or visceral processes responsible for metabolic reactions, there is no question that this area serves some inhibitory function. In contrast, the LH area appears to facilitate both ingestive and digestive processes. The pioneering work of Anand & Brobeck (1951) indicated that bilateral lesions in the area lateral to the VMH resulted in aphagia and eventually death in the operated rats. Follow-up work by Teitelbaum & Epstein (1962) indicated that if the LH-lesioned rats were given intragastric feeding, recovery of feeding occurred. They described a four-stage process of recovery. The severity of the aphagia and the length of the recovery was dependent upon the size of the LH lesion. Even though the lesioned rats recovered the ability to survive on dry food and water, they regulated their body weight at chronically lower-than-normal levels for an indefinite period. This occurred whether or not the rat was fed a highly palatable diet.

Powley & Keesey (1970) have shown that aphagia is a response to a chronic lowering in the rat's relative weight at the time the lesion is made. When LH lesions were made in animals that had been eating as much as they wanted, the rats remained aphagic until they had lost enough to reach about 80 per cent of normal weight. When another group of rats were starved pre-operatively, so that their weights were about 60 to 65 per cent of normal, they resumed eating after surgery until they gained weight up to the 80 per cent level. When a third group was force-fed so that they were heavier than normal, they showed a more prolonged period of aphagia. However, they resumed eating when a lower-than-normal body weight had been attained. Corbett & Keesey (1982) demonstrated that apahagia is not the only means by which LH lesions cause weight loss. Immediately after the lesion was made, the rats showed an increase in oxidative metabolism. Once the rats had lost enough weight to reach a lower-than-normal set point, their metabolism returned to normal. The results by Keesey's group led to the view that one of the functions of the LH is the regulation of the body's normal weight or set point. If the LH is damaged, the lack of eating and drinking is not a deficit in the ability to eat but rather a change in the body's set point to a new lower level. A reduction of eating is the way that the lesioned rat reduces its body weight to its new set point. Similarly, VMH-lesioned rats defend a new, higher body weight against dietary challenges in much the same way as LH-lesioned rats defend a lowered body weight. Some physiological and behavioural effects of LH lesions are presented in Fig. 4.3.

The most likely stimulus for regulation around a body weight set point is body fat. The LH and VMH work in a reciprocal fashion to determine a set point for adipose tissue. Because insulin is involved in the storage of energy in the fat cells, it may be involved in regulating the set point (Keesey & Powley, 1975). The effects of VMH lesions on insulin release is discussed in the preceding section 'The cephalic phase hypothesis'. Insulin secretion from the pancreas and levels circulating in the vascular system are proportional to adiposity (Woods, Decke & Vasselli, 1974). Insulin crosses the blood–brain barrier from the circulation via a saturable, receptor-mediated process of the brain (Baura *et al.*, 1993). There are insulin receptors in many regions of the brain that are in-

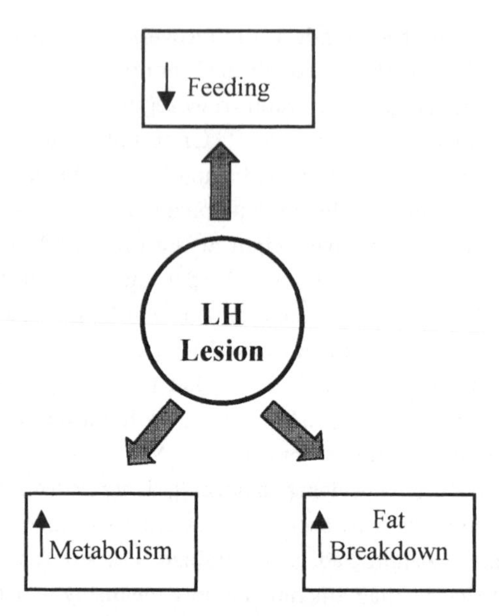

Fig. 4.3 The physiological and behavioural effects of lateral hypothalamus (LH) lesion.

volved in the regulation of food intake (Baskin *et al.*, 1988). These factors make insulin a good candidate for providing homeostatic feedback to the central nervous system about the state of peripheral fat stores. However, a second viable candidate is the recently identified protein hormone called 'Ob' or 'leptin', which is synthesised in and secreted from adipose cells (Caro *et al.*, 1996). The essential issues are that adiposity is regulated, circulating factors are involved and the overall system can be considered homeostatic. The consequences of VMH and LH lesions on body weight set point through its regulation of body fat are shown in Fig. 4.4.

General motivational and motor deficits arising from LH lesions

The effects of LH lesions are not specific to eating, but produce a wide range of other symptoms. Sensorimotor functions are also disturbed as a result of LH lesions. Intact rats respond to the presentation of stimuli in their vicinity with an orientation response. They track visual stimuli with their eyes, approach or withdraw from olfactory stimuli and react to tactile stimuli. Rats with bilateral LH lesions do not respond to any of these stimuli (Marshall, Turner & Teitelbaum, 1971). These rats also show a reduced responsiveness to the sensory properties of food and

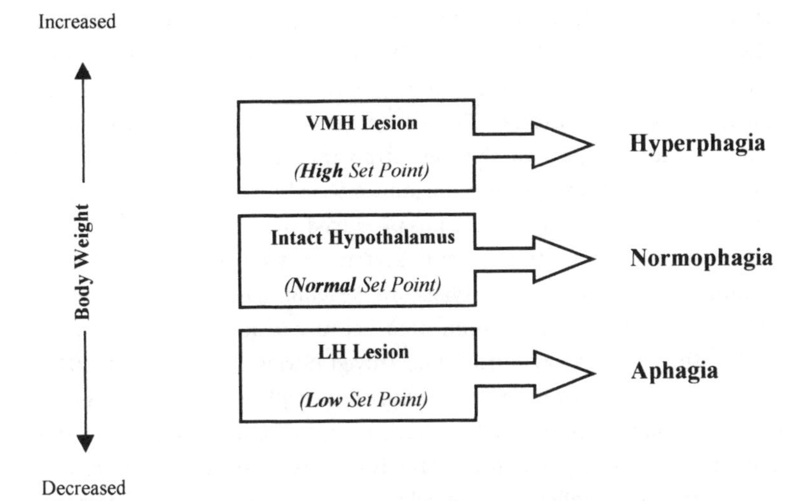

Fig. 4.4 Set point interpretation of body weight regulation implicating ventromedial hypothalamus (VMH) and lateral hypothalamus (LH) functions.

water, and in general, suffered disruptions of all motivated activities. Furthermore, Teitelbaum & Epstein (1962) found that recovering LH-lesioned rats can eat wet palatable foods even when they will not eat dry lab chow. In addition to these sensory deficits, LH-lesioned rats are impaired in their motor functions. During the immediate post-operative period the rats are quite inactive. They crouch in a hunched-over posture and fail to groom their bodies. They display a waxy flexibility and can be bent into peculiar positions and maintain them for prolonged periods. Control rats vigorously resist being placed in such postures. These sensorimotor functions initially disturbed by LH lesions show considerable recovery, but as with feeding and drinking, the time of recovery is variable. Interestingly enough, sensorimotor and appetitive functions generally recover concomitantly.

In his review of homeostatic origins of ingestive behaviour, Stricker (1990) argued that the LH lesions do not destroy feeding and drinking 'centres' in the manner proposed in the earlier explanations. He suggested that non-specific components of arousal in the performance of motivation of ingestive behaviours had been disrupted as a result of these lesions. The notion of non-specific arousal was central to Hebb's (1972) general theory of behaviour. LH-lesioned rats may be slow to respond, but when sufficiently stimulated, or given enough time, they can make many of the adaptive responses thought to be lost. Furthermore, Ungerstedt (1971) demonstrated that injection of the neurotoxin 6-hydroxydopamine (6-OHDA) directly into a large band of neural fibres known as the medial forebrain bundle, in which the LH resides, produces much of the LH syndrome. This neurotoxin destroys the cell body of neurons that are part of the neurotransmitter system involving *catecholamines,* which are protein hormones based on a single amino acid (monoamines). Catecholamines are produced in the synapses as transmitter substances of the nervous system as well as in the adrenal medulla, and two important types are *dopamine* (DA) and *noradrenaline* (NA). However, the story is complicated, because to be effective, 90–95 per cent of the DA pathway must be destroyed (Stricker & Zigmond, 1975) while electrolytic lesions in the LH cause only a 40–60 per cent depletion of DA (Oltmans & Harvey, 1972). Such complications indicate that the explanation of aphagia arising from LH lesions is not solely a matter of neurotransmitter deficiencies. Stricker's current position is that damage to the LH disrupts homeostasis, and striatal DA loss disrupts initiation and execution of feeding and related behaviour (Woods & Stricker, 1999). There has been considerable research done on the influence of neurotransmitters on feeding over the past few decades, and some of it is described in the section 'Physiological Analysis Revisited'.

If the effects of LH lesions were due to damage to fibres of passage travelling through or near this area, then aphagia should not result from injections of neurotoxins that selectively damage LH cells whilst sparing the fibres of passage (Clark *et al.*, 1991; Dunnett, Lane & Winn, 1985; Winn, Tarbuck & Dunnett, 1984). With this technique, damage to LH cells does produce a lasting decrease in food intake and body weight, and this is not associated with DA depletion due to damage to DA system pathways or with the sensorimotor deficits which are produced by damage to the DA systems. This series of experiments led Rolls (1994) to suggest that the LH does play a role in the control of feeding and body weight, but in a more intricate fashion than posited in Stellar's (1954) model. Details on Rolls's model is presented at the end of this chapter.

There is another effect of LH lesions that is of psychological significance, namely that food aversion is produced in addition to aphagia (Schallert & Wishaw, 1978; Stellar, Brooks & Mills, 1979; Teitelbaum & Epstein, 1962). The work of Cromwell & Berridge (1993) has refined this analysis to indicate that damage to other regions near the LH produce effects which separate aphagia from taste aversion. More specifically, damage restricted to the LH produced aphagia but not enhanced aversion when these rats were given the taste reactivity test. They used a method for assessing affective responses to tastes developed by Grill &

Norgren (1978) and which is described in the section 'Hedonic factors in taste reactions' in Chapter 5. The crucial region for aversion is located in a region lateral to the LH and into the nearby global pallidus or substantia innominata. The effects were not due to loss of 6-OHDA fibres but arose solely from cell loss of this region.

THE VMH SYNDROME AND THE ANALYSIS OF HUMAN OBESITY

In a classic paper, Schachter (1971), summarised his research which suggested parallels between the behaviour of obese humans and rats suffering from the VMH syndrome. Most of his studies involved comparisons between people who were of normal weight and those who were overweight. Although there is no reason to assume that the majority of obese humans suffer from undiagnosed hypothalamic injury, Schachter suggests that these groups differ in their disposition with respect to responsiveness to internal and external cues associated with food. The VMH syndrome indicates such dispositional differences in rats. Whereas individuals of normal weight and control rats regulate their intake according to the sensory properties (tastiness) of foods as well as internal homeostatic cues, this is not the case with the obese. The latter are mainly influenced by external environmental cues. With obese humans, cues such as time of day, taste of food and effort have greater control over their food intake than those for non-obese humans.

In the Schachter & Gross (1968) study, the speed of two clocks was altered so that one ran twice as fast as normal and the other ran half as fast as normal. The experiment was run just before dinner time (6 p.m.) and the subjects were left alone for 30 minutes, but the clocks in their rooms indicated that either one hour or 15 minutes had elapsed. The subjects who were in the room with the 'fast' clock thought it was after 6 p.m. when the experimenter returned to the room, while subjects with the 'slow' clock thought it was 5:20 p.m. While the subjects were occupied with a 'dummy' task (filling out a questionnaire), they were offered crackers as snacks. It was hypothesised that if the subjects were sensitive to external cues associated with eating, those in the 'fast' clock condition would eat more crackers than than those in the 'slow' clock condition. Associations between time of day and dinner would have a greater influence on externally oriented subjects. The results indicated that obese subjects in the 'fast' clock condition ate twice as many crackers as those in the 'slow' clock condition. In contrast, the normal subjects in the 'slow' clock condition ate more than those in the 'fast' clock condition. During the post-experiment interview, the latter subjects reported that they did not want to spoil their dinner by eating too many crackers. These results indicate that the obese and normal weight individuals respond to external cues associated with feeding in different ways.

Another external cue that influences eating behaviour is taste. Nisbett (1968) examined the effects of taste on the consumption of overweight, normal and underweight subjects when they were offered good- or bad-tasting ice cream. The results indicated that overweight subjects ate more of the good-tasting ice cream than normal or underweight subjects. These results support the externality hypothesis. However, the underweight and obese subjects ate more of the bad-tasting ice cream than the normal weight subjects. Schachter (1971) attributes this discrepancy to the fact that all three groups ate very little of the bad-tasting ice cream. Though significant, the group differences were small. Schachter & Rodin (1974) compared these effects with those indicating that VMH-lesioned rats show greater reduction of intake than non-lesioned rats when the diet is adulterated with quinine and enhanced intake when the food is sweetened. They suggest these similarities reflect a finicky reaction to the taste properties of food.

Self-reporting of hunger varied among obese but not among normal subjects (Nisbett, 1968). He suggested that the obese subjects are unresponsive to short-term changes in their nutritional state but show an exaggerated responsiveness to taste cues. Independent research by Cabanac & Duclaux (1970) showed that normal subjects find otherwise pleasant-tasting sweet solutions to be aversive when they sample them a few minutes after a gastric load of glucose. In contrast, obese subjects find the solutions somewhat more pleasant tasting than normal-weight individuals in the presence of the glucose load. It is interesting that the glucose load did not decrease the attractiveness of the sweet solution for the obese as it did for the normal weight subjects. A more recent study by Spiegel, Shrager & Stellar (1989) on the effects of gastric preloads, also found that obese subjects were more responsive to palatability and less responsive to deprivation than lean subjects.

Schachter's (1971) proposed relationship between obesity and sensitivity to external cues has stimulated much interest and research but many suggest that it cannnot be the whole story. In her review, Rodin (1981) noted that many studies have been unable to show that obese individuals are more responsive than normal weight individuals to external food (or non-food) cues. Externality is found in both lean and obese people, but not all obese individuals show it. Being overweight is not exclusively associated with externality. External stimuli as well as internal states can trigger motivational states. This is accomplished by external cues interacting with and creating internal states, which in turn influence eating. For example, both saliva and insulin secretion are increased in the presence of external cues that reliably predict the availability of food (Rodin, 1981).

In contrast to Schachter's (1971) emphasis on externality, Nisbett (1972) suggested that body weight set point, another concept derived from the rat studies, may be a factor which afflicts obese humans. If adipose tissue set point is defended by interactions between LH and VMH (Powley & Keesey, 1970), it is possible that different baselines in individuals are affected. The VMH–LH areas maintain whatever set point had been established by heredity and by nutritional conditions during childhood. This hypothesis implies that an individual could be obese and yet be at or even even below the set point that his or her system is trying to defend. If this were true, obese individuals who are at their set point should exhibit behaviour resembling those of normal weight individuals. Nisbett tested this prediction in the following manner. A large group of underweight, normal and overweight subjects were tested with good- and bad-tasting ice cream. Some individuals in the normal category were formerly overweight people who had lost weight. Thus these individuals along with underweight ones could be regarded as being below their set points. Overweight subjects were classified as being at their set points if they were 40 per cent or more overweight.

The results suggest that among subjects at or near their set points there is very little difference in the eating behaviour of obese and normal weight indivduals. Within each weight group the subjects below their set points were more responsive to the taste of ice cream than subjects at their set points. Thus, when obese individuals' degree of overweight is so great that they have allowed their weight to equal or exceed the set point, it appears that they behave like individuals of normal weight. However, when individuals are below their set point, they appear to behave like obese people even if they are of normal weight or underweight. Exciting though the results were, Nisbett's findings have not been consistently replicated in subsequent experiments. Rodin, Moskowitz & Bray (1976) studied the relationship between degree of obesity, nearness to set point for adipose tissue mass and responsiveness to taste. They found that the preference and intake of milkshakes of the moderately overweight subjects were more influenced by tastes that they found positive and negative than was the preference of normal weight or obese subjects. The ease of ingesting the taste substance also influenced preference and food intake of the moderately obese only. However, nearness to set point, operationally defined as weight stability for two years did not relate with taste responsiveness as predicted by Nisbett's (1972) theory.

Nisbett (1972) hypothesised that when the organism is in energy deficit, it behaves in a manner analogous to the VMH-lesioned rat. The behaviour of food-deprived rats and weight-watching individuals are essentially similar to that of rats with VMH lesions. Nisbett proposed that the VMH is attuned to the nutritional state, becoming less active when it moves into energy surplus. He cited Pfaff's (1969) finding that the nucleoli of VMH cells of starved rats are smaller than those of well-fed animals. Because nucleus size is correlated with cellular activity, Pfaff's results are consistent with Nisbett's hypothesis. His notions would be clearly substantiated if there was direct evidence indicating that the obese are below their body weight set points. Most of the arguments on this issue are indirect, but sound reasonable.

People who have lost weight have a tendency to gain it back again, and to return to the weight they were at before the loss occurred (Brownell & Stunkard, 1978). The body seems to resist loss of weight by every means available. Metabolic rate goes down in dieters so that less energy is wasted and less fat is broken down (Garrow *et al.*, 1978). This indicates that energy output as well as intake is affected. In addition, the processes that pull fat out of the blood, and converts it into stored fat molecules are accelerated. Persistent dieters are especially faced with this problem. Observations on people who attempt to put on weight and to maintain it also support the set point notion. Normal weight volunteer subjects who force themselves to overeat over a substantial period gain weight. But they gain less and less weight during this period, even though their caloric intake is high. This happens because energy output rises. More heat is generated, so that many of the excess calories are wasted rather than being stored as fat. The body resists weight gain just as fiercely as it resists weight loss (Keesey *et al.*, 1984; Keesey & Hirvonen, 1997).

Why do some individuals have a body weight set point that is set too high? In dealing with this question, Stricker (1978) divided obesities into two types. In one, overeating develops in response to non-regulatory influences such as good-tasting foods, or occurs in response to arousal or stress. People who become obese as a result of these conditions are fairly successful in losing weight by for example decreasing their consumption or reducing their source of stress or arousal. In the second type, the tendency to obesity may actually precede the overeating. Metabolic abnormalities may drive fuels out into the blood and into the fat cells, which turn them into more fat. That leaves the other cells starved for fuel, and chronic hunger could result. It could be possible that such people may not get fat because they overeat, but overeat because they are getting fat. Thus some people struggle against obesity as a life-long battle while others stay lean without even trying. Individual differences in body weight set point may be the important factor here. Because of this, the statistics on methods for decreasing obesity are not very encouraging (Van Itallie, 1984). Similarly, frequent cycles of food restriction ('dieting') and re-feeding in obese rats causes a fourfold increase in physiological food efficiency compared with obese rats that were fed on an *ad lib.* schedule (Brownell *et al.*, 1986). This finding is consistent with Brownell's (1982) observations that most of the weight lost as a result of treatment programmes is regained rather easily. Does this suggest that some obese people are fighting a lost cause?

Up until now we have been analysing obesity in terms of proximate mechanisms. If we attempt to analyse what possible function there is in the propensity to be obese, we may be able to answer the question posed in the preceding paragraph. Humans have evolved in an environment of unreliable food resources and scarcity of nutrients was a problem that had to be faced (Garn & Leonard, 1989). Thus, the ability to store energy for future use is adaptive. An individual of normal weight has about a month's supply of stored energy, and obesity may be a means to cope with famine. Because the energy reserves are not drawn upon by lack of adequate food, the storage system continues to add to already adequate reserves. In the past, the control of obesity was primarily environmental and humans may not have well-developed mechanisms for limiting fat storage beyond a minimum level, because nature usually limited the storage through famines. The reason why some people become obese while others do not has to do with genetic differences. Logue (1991) has questioned 'whether all the time, money and effort put into reducing mild to moderate (i.e. not life-threatening) obesity is worth the small number of pounds usually lost'. She predicts that interest in weight loss seems unlikely to diminish unless the present fashion image changes. This brings us to the topic of the next section, anorexia and failure of regulation.

BIOLOGICAL FACTORS IN ANOREXIA NERVOSA

When individuals consume inadequate amounts of food in the face of ready availability, they are said to exhibit anorexia or lack of appetite. Anorexia nervosa is an eating disorder which concerns 'the relentless pursuit of thinness through self-starvation, even unto death' (Bruch, 1973). It usually occurs in females – only five per cent of anorexics are males. This disorder also tends to be prevalent among white and middle or upper class people. Logue (1991) reviewed many studies indicating that the actual prevalence of anorexia, not just the reported prevalence, has increased during the past 25 to 35 years. A reduction of 25 per cent of original weight is considered a minimum value for diagnosis of anorexia nervosa. A second major symptom is amenorrhoea, absence of menstruation. If girls develop anorexia nervosa prior to reaching puberty, menarche is delayed. This disturbance is attributed to the lowered fat content of the body. A third major symptom is distorted attitude toward feeding that often includes denial of the need to eat, enjoyment in losing weight and a desired body image of extreme thinness. Anorexics have a different perception of their body size, compared with the impression that other people have of them. They insist that they are fat even when severely underweight (Bemis, 1978). This attitude causes them to engage in excessive exercise. Many patients exercise vigorously for many hours each day.

Clearly the anorexic is no longer maintaining a homeostatic energy balance and the problem is to understand the cause of her condition. A large number of explanations have been proposed and they include societal, psychological and physiological factors. From the 1960s to the present, thinness has been the fashionable image of young women. The majority of them feel that they need to be thinner in order to be attractive (Fallon & Rozin, 1985). Slenderness is often linked with femininity. This belief, which is prevalent among those in higher socioeconomic groups, motivates them to engage in the various activities resulting in drastic weight loss. Bemis (1978) asserts that the curtailment of intake is more often motivated by the desire of an extremely thin appearance than by genuine lack of hunger.

Another approach focuses on acquired behaviours of the individuals who develop the disorder. They may use this ailment as a means of gaining attention, as a denial of sexuality and as a tactic to deal with overdemanding parents (Bruch, 1973). Bruch depicted the mothers of anorexics as dominant and intrusive. The girl, in her struggle for self-determination, may refuse to eat as a means of asserting control over an important area of her life. This type of theoretical orientation is generally favoured by clinicians who based their ideas on observations of patients. However, experimental data supporting this position is sparse.

A third theoretical approach to anorexia nervosa emphasises the role of physiological variables. For example, failure to menstruate and anorexia appear to be related (Frisch, 1987). Some had reasoned that there may be a common physiological factor responsible for both the menstrual irregularities and anorexia nervosa. Since the hypothalamus is critically important in feeding behaviour and in hormonal functions, dysfunctions in that region may be an important determinant of anorexia (Garfinkel & Garner, 1982). Katz & Weiner (1975) had shown that the hypothalamic–pituitary–adrenal axis is activated among anorexics and that their adrenal cortex gland is unusually responsive. Cortisol levels are elevated during anorexia and appear to be an adaptation to starvation. This raises the question of whether this abnormal functioning is the cause or an effect of the lack of eating. Because most of the hypothalamic functions, including the functioning of the hormones that cause menstruation, return to normal once eating is resumed and weight is regained, it is possible that these abnormalities are an effect, rather than a cause, of the anorexia. We will return to the issues of oestrogens when dealing with 'activity-based anorexia' in the following section.

Because adrenal activity was found to be considerably higher in a group of anorexic patients than in control subjects, Walsh *et al.* (1978) suggested that anorexia may be an affective disorder. Cantwell *et al.* (1977) found some evidence of a family history of affective disorders among anorexics and observed that depressive symptoms were evident before and after recovery. Levels of a major metabolite of brain noradrenaline (norepinephrine) are lower in anorexic than in control patients (Halmi *et al.*, 1978). There remains the problem of discerning cause and effect that was encountered in our discussion in the previous paragraph. Is anorexia caused by the lowered catecholamine level or is the lowered level produced by anorexia? Until this problem is resolved, the hypothesis that anorexia is an affective disorder resulting from different standing levels of certain catecholamines has not been substantiated.

Activity-based anorexia

We had alluded to one of the salient characteristics of anorexics, their obsession with exercising as a means of reducing weight. The superimposition of exercise upon restricted feeding may precipitate anorexia nervosa (Aravich, Doerries & Rieg, 1994). Once precipitated, anorexia may be perpetuated by differential changes in specific brain opioids. An initial observation on rats by Spear & Hill (1962), followed-up by extensive analysis by Routtenberg & Kuznesof (1967), revealed a phenomenon known as activity-based anorexia (ABA). Although Bolles & de Lorge (1962) obtained similar effects in an experiment that was designed to test the effects of biological clocks, they chose to overlook the theoretical significance of the former. The ABA paradigm involves the imposition of a restricted feeding schedule on normal weight rats with free access to a running wheel. The behavioural consequences of this procedure are as follows. The rats progressively increased their activity levels while concomitantly decreasing their food intake. Thus their body weight decreased because food intake did not increase sufficiently to compensate for the energy expended in wheel-running. They ate less than control rats on the same feeding schedule who lived in standard cages. These animals also manifested abnormalities of neuroendocrine function that are found in the emaciated anorectic patient.

Later studies indicated a relationship between ABA and concentrations of an endogenous opiate found in the plasma and central regions of the nervous system (Doerries *et al.*, 1989). This opiate substance – β-endorphin – when injected into the VMH increased feeding activities in rats (Grandison & Guidotti, 1977). In the Doerries *et al.* (1989) study, the experimental group had access to food for 90 minutes daily and had free access to the running wheel throughout the experiment while the controls were housed and fed in their home cages. When the experimental rats had reached 70 per cent of their original body weight as a result of their running and decreased food intake, they were sacrificed (along with the controls) and radioimmunoassays were performed to ascertain the concentration of β-endorphins. The ABA animals showed elevated levels of this endorphin relative to the controls. Luby, Marrazzi & Sperti (1987), Marrazzi *et al.* (1990) and Marrazzi, Kinzie & Luby (1995) suggested that endorphins could be mediating the addictive-like behaviour of the patient with anorexia. During the period of initial dieting, endogenous opioids are released, then get the patient high on dieting and thus addicted to it. The opioids play a dual role in responding to starvation: (1) they increase food intake to correct it; and (2) they adapt for survival in the face of starvation until it is corrected, by down-regulating function to an essential minimum. If these responses become uncoupled, addiction can occur to one without the other. Among anorexics the addiction is to the elation and/or adaptation to starvation. Marrazzi's research provides support for the auto-addiction hypothesis as a result of three lines of evidence: endogenous opioid levels are induced by food deprivation among mice; anorexia patients show elevated total endogenous opioid activity in their cerebral spinal fluid; and the addiction characteristics of anorexia can be treated with narcotic antagonists.

As already mentioned one of the symptoms of anorexia nervosa is amenorrhoea. Watanabe, Hara & Ogawa (1992) tested young female rats under ABA conditions and found a concomitant disappearance of their oestrous cycle with augmentation of running activity. These findings suggest that amenorrhoea may reflect a physiological aspect of coping behaviour manifested by these rats. In general, the ABA paradigm models one of the perplexing aspects of the clinical syndrome, the ability of the patient to overwhealm homeostatic mechanisms for energy balance by a combination of restricted eating and excessive activity.

Recently, Dwyer & Boakes (1997) have argued against the notion that major weight loss is a direct consequence of the high level of activity in the wheels. They suggest that the progressive weight loss of the ABA effect occurs only when the running wheel is available at the time the rat is having to cope with the restricted feeding time. If the rat were given 90-minute access to food before the doors of the wheel are opened, food intake and weight loss are not significantly affected even though their running was considerable. Although the amount of running a rat undergoes is not important to ABA, *when* it runs it is critical. Dwyer & Boakes (1997) compared two groups of rats that were placed on a 90-minute feeding schedule. At first the rats were given access to the wheel, but later they were given restricted access. For ABA to occur, the rat had to run for the few hours immediately preceding the arrival of food. Those given access to the wheel many hours later did not experience the ABA effect. This led Boakes & Dwyer (1997) to conduct an experiment the results of which suggest that the key factor for ABA is the process of adjusting to an abnormally timed feeding schedule. Rats are normally nocturnal eaters, and the imposition of a 90-minute feeding schedule during the dark period has a minor impact on their body weights compared to the introduction of a 90-minute schedule of access to food during the day. If the standard ABA procedure is followed, but with feeding during the dark period, the ABA effect disappears. The implication of this analysis places constraints on the usefulness of the ABA model for explaining anorexia nervosa.

An evolutionary perspective on anorexia

The preceding discussion has been solely concerned with the proximate mechanisms of anorexia. An examination of this phenomenon from an evolutionary perspective may provide different insights on the phenomenon (Surbey, 1987). In many species females can optimise their lifetime reproductive success by suppressing reproduction when present conditions for the survival of the offspring are bleak and the future offers better prospects (Wasser & Barash, 1983). This can take the form of delay of sexual maturation or the prevention of ovulation. Female mammals under stress or in poor physical condition often show delayed or suppressed reproduction. In humans, starvation results in anovulation and amenorrhoea (Frisch & McArthur, 1974). Psychological stress also precipitates menstrual irregularities and amenorrhoea, and the self-starvation associated with anorexia nervosa. Amenorrhoea is viewed as secondary to starvation but the fact that it does not remit following weight gain indicates that this may not be so. Thus stress may trigger the amenorrhoea associated with anorexia nervosa. Starvation may only serve to reinforce the suppression of menstruation once it is initiated by stressful conditions.

Anorexia nervosa affects mainly early- rather than late-maturing girls. As the age of menarche has decreased, the incidence of anorexia nervosa has increased because of the emotional turbulence at puberty. Early maturers appear not to have the necessary knowledge and skill to deal with menarche and may feel fear and disgust at the physiological signals of their approaching womanhood. Early sexual maturity has been correlated with early sexual activity, higher rates of premarital pregnancy, early marriage and early first births (Presser, 1978; Udry & Cliquet, 1982). Like early maturers, anorexics express ambivalence and fear of sexual development (Bruch, 1973). Loss of secondary sexual characteristics through dieting may be seen as a method of turning back the clock.

Surbey (1987) hypothesised that the developmental pathways of early and late maturers reflect different strategies, the type adopted depending upon environmental conditions. The development of anorexia nervosa may be a means of altering the developmental trajectory from that of an early maturer to that of a late maturer. Through the reduction of sexual fat, the anorexic girl attracts fewer males and thus needs to deal less with sexual advances. Self-starvation reduces libido and sexual frustration is thereby avoided. This leaves the anorexic more time to pursue academic success and a satisfying career. Feierman (1987) described anorexia nervosa as a method which allows a female to determine time of conception. It also allows her to develop the skills for successful reproduction. According to the evolutionary perspective, anorexia nervosa is not the result of selection pressures but reflects a fertility control mechanism adaptive in our evolutionary past.

PHYSIOLOGICAL ANALYSIS REVISITED

VMH syndrome

Much of the earlier discussion in this chapter on feeding mechanisms was based on the seminal work of Stellar (1954) who suggested that the

VMH operates like a 'satiety centre'. The two hypothalamic centres he described were part of a negative feedback loop that was completed by the impact of correction on the internal and external environment. He assumed the links between hypothalamic excitatory centre and the final common pathway for behaviour is direct, and that motivated behaviour is switched on and off by the changes in the motivational state. However, subsequent research has produced data that require more intricate interpretations. The earlier work was guided by the assumption that if destruction of a particular part of the brain results in loss of a specific function, then that area is essential to the ability to manifest that function. However, lesions impose relatively crude assaults on the region and their effects may reflect the functional properties of the brain than about structure–function (brain–behaviour) relations (Staddon, 1983, pp. 184–185). Carefully placed lesions in a given region may also damage other nearby structures. Because functionally different structures overlap physically, more than one functional subsystem may be damaged by a lesion. Lesions of the VMH are not the only lesions that can induce hyperphagia. In fact, lesions of the paraventricular nucleus, which is located just ahead of the VMH may be more effective in eliminating satiety than lesions of the latter (Stellar & Stellar, 1985). In retrospect, it is possible that earlier studies may have limited the focus to a more restricted region of the brain. There are many connecting neural fibres that may be interrupted by these lesions, and a host of interconnecting neuronal pathways is responsible for 'VMH effects'. The hypothalmus is but one of the major integrators of external stimuli, internal sensory stimuli and activity from other parts of the brain.

In summarising his views on VMH syndrome, Booth (1994) pointed out an irony in that this work diverted social psychologists into the biology of obesity as a result of the research stemming from rats which later was found to be misleading. Schachter (1968) got inspiration for the 'externality' theory of human obesity from the emotionality of VMH obese rats (Schachter & Rodin, 1974). Yet Graff & Stellar (1962) previously showed that emotional reactivity of VMH rats is not related to their obesity, a finding that was substantiated with a different method by Hallonquist & Brandes (1981a). The former demonstrated that the two symptoms arose from damage in different parts of the VMH area.

Neurotransmitter and modulators

Advances in the influence of neurochemicals on feeding behaviour are considerable. They have been possible as a result of two principal methods; one involves studying the effects of specific neurotransmitters, antagonists, or agonists introduced into discrete parts of the brain with a cannula, and the other involves direct measurement of the content or release of transmitters in certain defined brain areas. The use of specific chemical toxins such as 6-OHDA to destroy catecholamine-containing cells (e.g. Ungerstedt, 1971) is very effective because the method produces more selectivity than the earlier tissue-producing lesions. However, this method suffers from similar problems of interpretation because even if we can accurately describe the functional deficits, it is still difficult to ascribe that function to the damaged region. Disruption in one site may cause dysfunction at distal sites that may be involved in that function. Nevertheless, we may still link feeding behaviour to various neurotransmitter systems. Current views of synaptic function include the concept of more than one transmitter (colocationization-release) with modulation of the action of the 'primary' transmitter either by the cotransmitter or by other actions on the postsynaptic membrane (Rowland, Li & Morien, 1996).

The direct measurement methods hold great promise because of their sensitivity. Early studies used whole-tissue measurements of neurotransmitters or their metabolites as measures of synaptic release or turnover. More recently, use of microdialysis with relatively small probes inserted into discrete brain regions allows an on-line measurement of extracellular transmitter concentrations. This method is the basis of a 'chemoencephalogram' (CEG) which provides excellent temporal and spatial resolution. The main limitation of this method is that researchers are usually limited to a single location and a single transmitter. Some of the neurotransmitters involved in the regulation of feeding are listed in Fig. 4.5.

Noradrenaline (NA) also known as norepinephrine (NE), was the first neurotransmitter linked to changes in feeding behaviour (Grossman, 1960). Normally, NA is used especially by neurons that have cell bodies which are located in the brainstem nucleus known as the locus coeruleus. Nevertheless, injections of NA into the hypothalamus have appetite-stimulating effects if introduced at sites such as the paraventricular nucleus (PVN) and appetite-reducing effects at sites such as the perifornical area (Leibowitz, 1986). CEG studies are consistent with such results. The extracellular concentrations of NA, detected by push–pull cannula or microdialysis in the PVN, increase with hunger-related treatments (Stanley *et al.*, 1989).

The PVN is also the site where other agents such as *neuropeptide Y* (NPY) and *galanin*, another amino acid peptide, are concentrated.

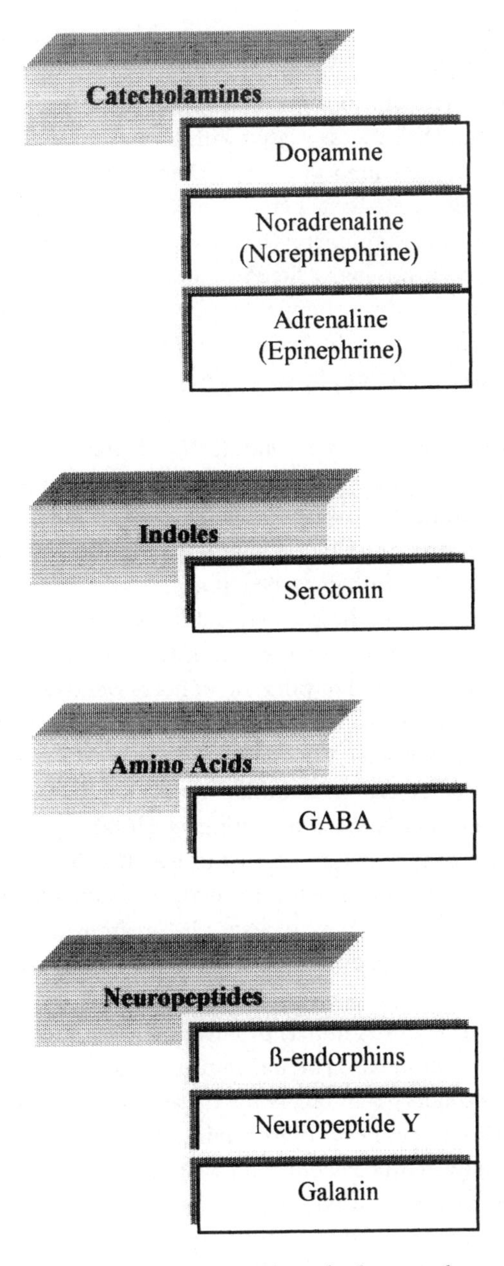

Fig. 4.5 Some neurotransmitters in the central nervous system involved in feeding.

Injections of NPY into this region causes sated rats to eat large amounts of food (Clark *et al.*, 1984). In addition, the extracellular and possibly the whole tissue concentrations of NPY show regionally specific increases in physiological conditions that are associated with increased food intake (Beck *et al.*, 1990). NPY plays an important role in the regulation of energy balance, and is chemically related to pancreatic peptides. The synthesis of NPY is associated with negative energy balance. The other peptide – galanin – has high concentrations in the PVN, and injections into this region stimulate food intake in rats (Tempel, Leibowitz & Leibowitz, 1988). The similarities and differences in behaviours induced by NPY and galanin may be due to interactions between NA, galanin and NPY (Kyrkouli *et al.*, 1990).

Another neurotransmitter – serotonin (5HT) – is implicated as a factor that inhibits or decreases feeding behaviour and thus satiety (Blundell & Hill, 1992; Blundell & Latham, 1979). Drugs such as dexenfenfluramine (DFEN) and fluoxetine, that function as 5HT agonists are potent anorectic agents (Blundell & Latham, 1995). Injection of 5HT receptor antagonists blocks the actions of these drugs, and supports the view that 5HT is critical for their action. However, the sites of action of these agents has not been determined and it is likely that they affect several regions of the brain (Li, Spector & Rowland, 1994). Li & Rowland (1994) suggested that DFEN and hormonal substance, CCK (see section 'Peripheral mechanisms of feeding') share some common similarities in their behavioural effects. Both seem to act on satiation, attenuating food intake by decreasing meal size without change in the latency to eat. This led Cooper, Dourish & Clifton (1992) to propose that these two factors interact.

The neurotransmitter that has attracted the greatest interest in the motivation of feeding is dopamine (DA). Activation of DA systems, as quantified by electrophysiological measures, microdialysis or voltrammetric measures, is triggered in animals by encounters with food, sex, drugs of abuse, electrical stimulation of the brain area that supports self-stimulation and by secondary reinforcers for these incentives (see the review by Berridge & Robinson, 1998). These findings suggest that DA has general effects on motivated behaviour, and that it is not specific to feeding. These studies involve pharmacological blockade of DA receptors in rats wherein DA antagonists reduce reward-directed and consummatory behaviour by slowing the rate and amount of voluntary eating of food pellets (Wise & Raptis, 1986). Dramatic effects are also produced by extensive DA depletion caused by intracranial application of catecholamine-selective neurotoxins such as 6-OHDA, as discussed earlier. After extensive destruction of ascending DA neurons in an area called the substantia nigra, rats become oblivious to food and many other rewards. They become aphagic and adipsic (i.e. will not drink) after 6-OHDA lesions, and will starve to death unless nourished artificially, even though food is available. Such rats retain the ability to walk, chew, swallow, perform other movements required for eating, but fail to employ them to gain access to food. Although DA depletion in the substantia nigra causes severe aphagia, it does not alter taste reactivity in response to oral infusion (Berridge, Vernier & Robinson, 1989). This intriguing set of results led other researchers to study the relationship between DA and different components of the feeding system with a different method.

Studies of the relationship between activity in areas rich in DA, such as the nucleus accumbens, with feeding indicate that this behaviour is not associated with increased utilisation of the neurotransmitter, but more with stimuli signalling its presence (Blackburn *et al.*, 1989). These results led the experimenters to conclude that 'dopamine systems are more importantly involved in *preparatory* than in *consummatory* feeding behaviours'. On presenting a cue predictive of food, normal rats orient towards the cue and explore the feeding tray. A DA antagonist drug such as pimozide decreases the vigour of such behaviour but has little effect upon consummatory behaviour (Blackburn, Phillips & Fibiger, 1987; Phillips, Pfaus & Blaha, 1991), a finding consonant with those reported by Salamone *et al.* (1991). Such results are consistent with Berridge *et al.*'s (1989) theoretical perspective which proposes that pimozide lowers the incentive value (salience) of food at a distance from the animal. These authors propose a general model in which 'mesostriatal dopamine neurons belong to a system that assign salience or motivational significance to the perception of intrinsically neutral events'. Such a notion suggests the importance of psychological factors in the feeding system. It is not merely a matter of physiological feedback. Feedforward mechanisms involving anticipatory processes such as learning are integral in the analysis of proximate mechanisms of feeding.

SELECTED GENERAL MODELS OF FEEDING

Stricker (1990)

There are many excellent recent models of feeding reviewed by Stellar (1990), so I will not revisit them with you. Instead, I have selected a few that integrate many of the issues that we have examined in the preceding sections of this chapter. Stricker (1990) has proposed a comprehensive model that explains feeding in terms of the multiple factors reviewed in the preceding sections. He views the motivation to eat as part of a general arousal mechanism, and proposes central excitatory and central inhibitory systems in the brain that act in a reciprocal fashion comparable to that proposed in earlier models. When a decrease in calories coming from the intestine and liver occurs, metabolic signals and sensory input from taste and smell pathways arouse the central excitatory system. Activity of the catecholamines arouses the organisms generally, and produces the feeding responses. The resulting calories from the intestine and liver along with insulin release contribute to satiation by inhibiting the metabolic arousal signals. The central inhibitory system mediated by the neurotransmitter 5HT and the central excitatory system mediated by catecholamines act reciprocally upon each other. Stricker's ideas are useful because they explain the starting and stopping of eating and integrate information on peripheral, metabolic, endocrine and sensory input to the brain. The integration of such an approach with the recent developments on the role of the DA system on incentive motivation should provide us with a better understanding of proximate mechanisms of feeding. A schematic diagram of Stricker's (1990) model of feeding is presented in Fig 4.6.

Rolls (1994)

In an earlier section, the argument that the hypothalamus plays a role in the control of feeding and food intake put forward by Rolls (1994) was discussed. In contrast to the earlier models, he specifies the manner in which this structure articulates with the brain systems that perform sensory analysis, such as the olfactory and taste pathways, other brain systems that are involved in learning about foods such as the amygdala and orbitofrontal cortex, as well as brain systems that are involved in the initiation of feeding such as the striatum. In his research with monkeys, Rolls (1981a,b) found a population of neurons in the lateral hypothalmus and another nearby structure – substantia innominata – which are related to feeding. These neurons respond to the taste and/or sight of food, but only if the monkeys are hungry (Burton, Rolls & Mora, 1976). Thus neuronal responses to food, which occur in the hypothalamus, depend upon the motivational state of the animal. The signals that reflect this state include gastric distension and the presence of food in the duodenum (Gibbs, Maddison & Rolls, 1981), as well as the glucose level in the plasma (Le Magnen, 1992). The responses of these neurons may reflect the rewarding value or the pleasantness of food since stimu-

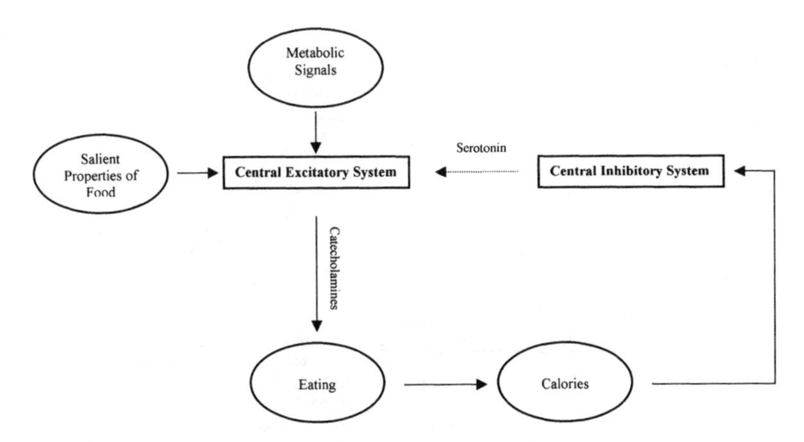

Fig. 4.6 Stricker's (1990) model of multiple factors that control feeding.

lation in this region can mimic the reward value of food (Rolls, Burton & Mora, 1980). This 'suggests that animals work to obtain activation of these neurons by food, and that this is what makes food rewarding' (Rolls, 1994, p. 19). These neurons may also be on a route for forebrain-decoded stimuli, such as the sight of food, to produce autonomic responses, such as salivation and insulin release. This aspect of Roll's theory is congruent with the *cephalic phase hypothesis.* He also suggests that a route for information about visual stimuli relevant to food is provided by temporal lobe structures such as the inferior temporal visual cortex and amygdala. The orbitofrontal cortex contains a population of neurons which is important in correcting feeding responses as a result of learning. Its functions are closely related to those of the amygdala.

The striatum is a structure belonging to the basal ganglia which is the terminal structure of the dopaminergic nigrostriatal pathway. Many of the brain systems implicated in the control of feeding, such as the amygdala and the orbitofrontal cortex, have projections to the striatum, which provides a route for these brain systems to lead to feeding responses (Mogenson, Jones & Yim, 1980; Rolls, 1984). Rolls (1994) suggested that the striatum may be crucial in bringing together food reward signals with information from many parts of the cerebral cortex, which are required for the initiation of actions such as eating. This explains the earlier material we had encountered on damage to the nigrostriatal bundle, which depletes the striatum of DA, and thereby produces aphagia and adipsia associated with a sensori-motor disturbance in the rat. The connections of the taste, olfactory and visual pathways described in Rolls's (1994) model of feeding are shown in Fig. 4.7.

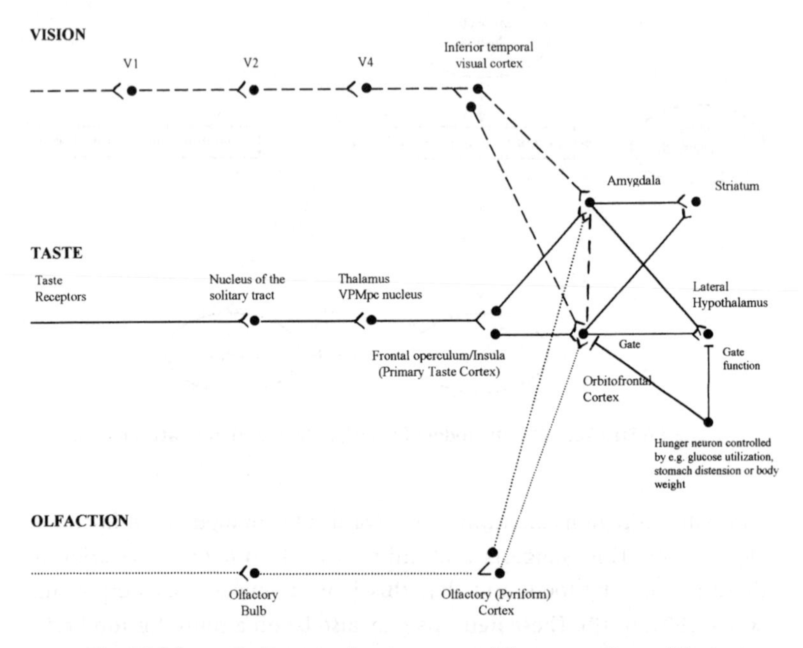

Fig. 4.7 Schematic diagram showing some of the connections of the taste, olfactory and visual pathways in the brain as proposed in Rolls's (1999) model of feeding. (Redrawn from Rolls, 1999, copyright 1999, by permission of Oxford University Press.)

Bernardis & Bellinger (1996)

Bernardis & Bellinger (1996) proposed a 'gestalt view' of feeding in which the LH figures in a food intake and body weight regulatory scheme that is related to the rate of energy production within this region. Because feeding stops long before the nutrients are absorbed, these authors hypothesised that elements of the LH sense pre-absorptive satiety in advance of intestinal absorption. Pre-absorptive signals are generated when food contacts the surface of the mouth and the gastro-intestinal tract, whence it triggers neuronal messages. The ultimate feeding stimulus is a drop in power or energy production within structures that regulate the general metabolism of the body. Bernardis and Bellinger suggested that the LH plays a role in the integration of food intake and physiological parameters of body weight control through its connection with the DMNV as well as the VMH. The role of the DMNV was described in the section 'The cephalic phase hypothesis'.

In contrast to theories that propose that a specific metabolic factor could control the onset of feeding and satiety, Bernardis & Bellinger (1996) suggested a model that focuses on feedback signals from power or energy production signals from many sources. They agreed that the previous theories are all valid but that each alone accounted for only part of the overall integrative scheme. It is for this reason that the authors adopted a gestalt view of feeding.

Raible (1995)

Although she did not develop a formal model, Raible (1995) made an attempt to present a coherent analysis of feeding mechanisms through the integration of causal factors at many levels. Raible organised her material in terms of control functions of three interacting regulatory systems. The emergency system consists of physiological and behavioural mechanisms that are activated when the organism experiences prolonged food restriction. The functions of this system are to disinhibit mechanisms of fuel conservation. The short-term system is involved in the initiation and termination of daily feedings as well as food preferences, a topic which will be discussed in detail in Chapter 5. The long-term system involves mechanisms that regulate fat deposits and may include mechanisms related to foraging strategies for food. Its main function is the regulation of energy stores.

The role of the emergency system is to increase food seeking under adverse deprivation conditions. The primary source of energy for the central nervous system (CNS) is glucose. During the absorptive phase of the feeding cycle, glucose is readily derived from carbohydrates, and during the fasting phase, glucose is obtained from endogenous energy stores such as fats and glycogen released from the liver. For these reasons, Raible suggests that the glucoreceptors comprise an important element of the emergency regulatory system. Glucoreceptors in the CNS that respond to glucose availability may be the 'inhibitors' and 'disinhibitors' of such behaviour. Earlier, we reviewed studies indicating that injections of NPY in the PVN result in increased food intake. More precise methods of infusing very small quantitities of NPY through very fine cannulae found that the most effective site was in the midlateral hypothalamus (Stanley *et al.*, 1993). S. Leibowitz's group (Leibowitz, 1978; Leibowitz & Brown, 1980; Stanley *et al.*, 1985) demonstrated that the increase in food intake was mainly due to carbohydrate ingestion, a finding that led Raible to suggest the PVN as part of a system that regulates glucose levels. Raible proposed that the LH and/or VMH glucoreceptors are located on neurons that synapse upon PVN neurons, and provide a means by which changes in glucose level produce changes in carbohydrate ingestion. Thus, the emergency system may specifically act to increase carbohydrate ingestion and thereby facilitate glucose utilisation.

The primary role of the short-term system is the regulation of meal patterns and food intake over a one- to two-day period. Psychological and environmental factors have a greater effect on this system than they do on the emergency system. The physiological bases of the short-term regulation of feeding involves noradrenergic and serotonergic neurotransmitter systems that act in opposition to each other. Activation of the noradrenergic system specifically increases carbohydrate ingestion whilst activation of the serotonergic system decreases it. Leibowitz, Weiss & Shor-Posner (1988) proposed that such an interaction occurs within the medial hypothalamus and PVN. Signals for 'nutrient monitors' in the liver and in the CNS are translated into changes in NA and 5HT release that act through the PVN and other brain areas to alter food intake. Hormones such as oestrogen influence feeding because of mediation through neurotransmitter systems. The inhibiting influence of oestrogen on feeding in rats may be mediated through its interaction with the serotonergic system.

The primary role of the long-term system is the regulation of the animal's buffer against privation. This can be achieved through the regulation of the animal's fat deposits as well as through food-hoarding behaviour. Mechanisms of the long-term system may alter food intake indirectly through changes in the body weight 'set point' through modifying the palatability of food or the animal's willingness to work for food. An animal may overeat or hoard food in excess of its physiological requirement because it is easy to obtain or because it is highly palatable. In her summary of physiological mechanisms of feeding behaviour, Raible made a bold sweeping statement to the effect that every neurotransmitter, sex steroid and peptide that influences food intake does so by acting in the PVN, VMH and/or LH. The coherence and elegance of this analysis is very attractive even though alternative models may indicate the need for greater complexity.

Ramsay, Seely, Bolles & Woods (1996)

These authors offered a reinterpretation of the depletion–repletion models of ingestive behaviour that we have examined in the preceding sections. It summarises many of the ideas of R.C. Bolles, a major figure in motivational theory, and those of S.C. Woods, who has made substantial contributions to the physiology of feeding mechanisms. This paper (Ramsay *et al.*, 1996) was a tribute to Bolles who contributed to its early development but died before the project was completed. These authors argued against the premise of predominant models of feeding – that feeding is initiated when the body becomes depleted in some crucial parameter related to food, such as blood glucose or stored fat. Negative feedback loops are but one of several strategies by which important parameters can be regulated. There is, however, a major drawback to this explanation. An error signal must be generated to activate the corrective response such that the animal must experience a perturbation of an important physiological parameter before corrective measures are initiated. Thus, these authors suggest feedforward regulation as a mechanism to address this limitation. A correlate of an impending and imminent perturbation of a regulated parameter reliably precedes the actual perturbation. Animals stop eating long before ingested nutrients enter the circulation in large quantitities, hence pre-absorptive feedforward signals are suggested as a major limiter of meal size. They cause the animal to stop eating well before the meal being consumed can have a major impact within the body. Hormones and other signals secreted when food begins entering the intestine are hypothesised to mediate this feedforward satiety system.

Although a hard-wire reflexive feedforward mechanism is effective in defending homeostatic equilibrium, Ramsay *et al.* (1996) suggested that it is not enough, and that learning is a special adaptive mechanism that allows for regulation in the absence of error signals. Through experience with particular foods and situations, animals learn to activate appropriate responses in anticipation of upcoming events. We have seen in the section on the *cephalic phase hypothesis* how a food-related increase of blood glucose can be anticipated by an animal through the existence of a sugar-signalling sweet taste, and how the postprandial increment of blood glucose is reduced through the release of insulin. This early secreted cephalic insulin is mediated by a neural connection from the brain to the pancreas (Powley, 1977; Woods, 1991; Woods & Kulkosky, 1976), and it is modifiable by learning. Thus learning provides valuable flexibility to homeostatic regulation. Predictive cues can entrain corrective responses so that they can be elicited before an actual homeostatic disturbance. Evolution may have shaped some classes of cues, such as tastes and smells, to become more easily associated with the post-ingestive consequences of food (Garcia & Koelling, 1966). If an animal lives in a sufficiently predictable environment, it can learn to avoid homeostatic disturbances by using anticipatory compensatory responses. The elicitation of learned responses that circumvent homeostatic disruption is a special case of a feedforward system. This analysis explains an obvious tendency among humans to eat not because of hunger but to avoid this state. We also eat because of pleasure associated with the specific tastes and smells and to avoid the displeasure experienced whenever we encounter other specific tastes and smells. This topic is treated in further detail in Chapter 5 'Food selection'.

SUMMARY

Feeding is a relatively simple behaviour influenced by 'a complex array of stimuli and and situational variables, the details of which remain to be fully understood' (Woods & Stricker, 1999). It is regulated by control systems incorporating feedback as well as feedforward mechanisms. This system provides the organism with the means to deal with homeostatic perturbations, and initiate anticipatory responses to them. Through the central nervous system the body is able to regulate how much should be eaten on the basis of how much has been eaten. Because of the complexity of the system, there are many models of feeding, and most implicate regions of the hypothalamus and related structures such as the paraventricular nucleus. Cells in the ventromedial and lateral areas are implicated in excitatory and inhibitory functions, respectively. Neurons in these regions also are activated by different neurotransmitter substances. Stimulation by catecholamines and neuropeptide Y excite feeding, whilst the release of serotonin inhibits the feeding reponse.

In general, signals about ingested food and body adiposity are integrated with information about the taste of the food, the memory of that food, the experience of previous meals, the competition of other desires and aspects of the environment (Woods & Stricker, 1999). The consequences of human feeding dysfunctions, such as obesity and anorexia, are analysed from the perspective of such biological and psychological factors. In human history the control of obesity was primarily environmental, and there may not have been well-developed mechanisms for limiting fat storage because nature limited the storage through famines. Genetic differences in body weight set point mediated by metabolic activity and disposition may underlie why some people become obese and others do not. Like obesity, anorexia is an eating disorder that reflects social, psychological and physiological factors.

5

Food selection

Most of the literature in psychology and physiology that deals with feeding in animals and humans focuses on hunger or how much is eaten (the regulation of food intake) rather than what is eaten (the selection of foods). Motivation textbooks rarely elaborate on the determinants of food selection, and if they do, such discussions occur in the section on hunger. The fact that food habits are used in the naming of higher-level taxonomic groups (e.g. carnivores, herbivores or omnivores, etc.) supports the notion that food selection is a major force in evolution. The prominence of food selections in our daily life is also evident amongst humans. If animals and humans are to survive and reproduce in their environments, they must find and eat foods that provide all the nutrients necessary for self-maintenance and reproduction, as well as to avoid eating lethal amounts of toxic plants or animals that they encounter.

This chapter describes and discusses the mechanisms that explain how animals and humans manage to discriminate foods containing needed nutrients from ingestible sources that are either valueless or dangerous to eat. The other major issue concerns the tendency amongst humans to be selective in their acceptance of a small subset of many edible items that are available to them. Explanations of these two major issues will be accomplished by a discussion of built-in 'hardware and software programmes' cognitive or learning processes and socio-cultural factors (mainly in humans) which are responsible for the choice of foods. Rozin & Schulkin (1990) specified seven aspects of the food selection 'cycle': (1) arousal of an interest in food or specific food; (2) search for appropriate foods; (3) recognition of such foods; (4) a 'decision' to consume such foods; (5) capture; (6) processing; and (7) ingestion. Research on the issue of arousal has focused upon the study of a few major nutritional categories such as food calories in investigations of hunger, water in thirst and sodium in salt hunger. The issue concerning search, capture and processing are studied by those working in the domain of natural history and foraging behaviour. Within psychology, experimental work has focused on the scientific study of food selection, and its inception was due to the efforts of Richter (1943). The main question of concern to psychologists in this field has been ontogenetic – to what extent is the choice innate or learned?

A major subproblem of food selection concerns choice of foods with different nutritional composition. Any consideration of food selection must take into account fundamental aspects of nutrition and physiology. There are thousands of different molecules that are necessary nutrients. Most are synthesised within the organism, including all the proteins. For chemicals that cannot be synthesised at a rate sufficient to maintain life, ingestive behaviour is required. There are differences in the range or level in which certain nutrients are regulated in the body. For example, body levels of sodium and water are tightly regulated and vary slightly. For other nutrients, such as vitamins, there is simply a lower limit. Excesses of such nutrients are stored, metabolised or excreted, but unlike the situation with sodium and water, there is no behavioural satiety effect. For such non-regulatory systems, a nutrient-specific internal state that signals deficiency is lacking; instead, a general state of malaise motivates a change in food selection. Detailed discussion on these issues is contained in a later section of this chapter dealing with the effects of vitamin B_1 or thiamine deficiency.

There are fundamental nutrient categories, such as protein, carbodydrate, fat, water and sodium. These categories are not equally identified by receptor systems because of the difficulties in designing a single receptor for all of them. For example, it is easy to design a sodium receptor, and it is widespread in animals. In contrast, it might even be possible to design a single receptor that would detect the density of food calories in a potential food. In addition to the issue of nutrient categories and receptor system, there are two fundamental components to food selection (Rozin, 1976). One is an internal state detector, which arouses appropriate food search based on a nutritional state, and the other is a food detector or recognition device, which identifies appropriate objects in the environment. In the ensuing sections of this chapter material is presented that indicates the fundamental difference between the nutrient categories of sodium/water and the others with respect to these detectors.

MECHANISMS OF FOOD SELECTION

Given the fact that animals, including humans, are selective about what they eat (there are many edible items passed over, especially by humans), the major question concerns the mechanisms responsible for such selectivity. At the most basic level, all vertebrates inherit sensory and central neural structures that cause specific reactions to the cue or sensory properties of certain food flavours. The first line of contact with food is the tongue, which contains specialised nerve receptors called taste buds or receptors that respond specifically to certain chemical substances. The sensation of taste occurs when chemicals stimulate taste receptors on the tongue and other parts of the oropharynx. Taste stimuli are separated into a small number of 'primary' tastes: sweet; salty; bitter; and sour. A wide range of sapid (tasty) stimuli applied to the tongue produces a profile of either neural or behavioural activity elicited by each chemical. For example, the activity evoked by about 50 single neurons through application of saline to a human tongue constitutes the response profile for sodium. This profile is different from those produced by chemicals that elicit the three other basic tastes. In other words, there are 'sweet-sensitive neurons' as well as 'salt-sensitive', 'bitter-sensitive' and 'sour-sensitive' ones. The evidence of four basic taste qualities emerged during the end of the nineteenth century (Öhrwall, 1901), and this position has been supported by modern research (McBurney & Bartoshuk, 1973).

Hedonic factors in taste reactions

Early research by Young (1948; 1961) had documented the influence of stimulus properties of foods on food choice, and led to the concept of palatability as 'the immediate affective reaction (liking or disliking) of an organism which occurs when a food comes into contact with the head receptors' (Young, 1948, p. 301). The notion of hedonic events influencing reactions to food sources can be traced to Darwin's (1872) examples of facial expressions that express pleasure or disgust to various tastes. In recent years, facial expressions have been studied extensively in objective terms, and assigned to different basic emotions (Ekman *et al.*, 1987). The analysis of facial expression to taste stimuli in human infants during the first few post-partum hours reveals distinctive negative responses to bitter- and sour-tasting substances and positive responses to sweet stimuli (Steiner, 1979; Rosenstein & Oster, 1988). Newborn infants do not show any distinctive facial response to salty stimuli, and salt

Fig. 5.1 Taste-reactivity pattern of the rat revealed by ingestive response elicited by intra-oral infusion of sweet-tasting solution. Ingestive reponses include rhythmic mouth movements, lateral tongue profusions and paw licking. (Drawing by Toni Ignacio based on material from Grill & Berridge, 1985.)

preference emerges at approximately four months of age. This reflects the influence of postnatal maturation of central or peripheral mechanisms on salt taste perception. Animal data also indicate that rats display characteristic oral–facial and other gestures to bitter and sweet stimuli analgous to those expressed by humans (Grill & Norgren 1978). In addition, sodium-deficient rats express greater oral–facial positive reactions to sodium tastes than non-deficient ones (Berridge *et al.*, 1984). Details on this method of assessing hedonic changes will be presented in the following section on hunger/appetite and salt. Fig. 5.1 depicts the taste-reactivity pattern of the rat to a sweet-tasting solution, whilst Fig. 5.2 reveals its reactivity pattern to a bitter-tasting solution.

Fig. 5.2 Taste reactivity pattern of the rats revealed by ingestive reponse elicited by intra-oral infusion of bitter-tasting solution. Ingestive responses include gapes, chin rubs, head shakes, paw wipes, forelimb flailing and locomotion (not shown). (Drawing by Toni Ignacio based on material from Grill & Berridge, 1985.)

The preceding discussion suggests that humans and other animals are hardwired to react to sweet- and salty-tasting foods with positive affect and thus show a high intake of such items. Conversely, sour- and bitter-tasting foods elicit a negative reaction and we tend to avoid ingesting substances containing a high level of substances causing such sensations that evoke such reactions. Rozin (1976) suggested that there is an innate attraction to sweet tastes (associated in the natural environment with fruits) and an avoidance of bitter tastes (associated in the natural environment with the presence of toxins). The widespread use and popularity of sweeteners can be traced to the biological desire for sweets (Rozin, 1982). A functional level of explanation would focus on the beneficial consequences of the liking of and ingestion of sodium and sugars. These substances are essential to the functioning of our bodies, and we do not have to learn to like foods that contain them. Rejection and avoidance of sour- and bitter-tasting foods is an adaptive defensive reaction. Sour tastes are associated with substances containing a high level of acid and rotting/fermenting material. In plants, bitter tastes are produced by chemicals with an alkaloid content, which in sufficient concentrations, could have toxic effects. Plants that have tissues which contain high amounts of quinine, mustard, nicotine, caffeine and other bitter-tasting substances, are less likely to be ingested by animals. Crawley (1983), Janzen (1977), and Rosenthal, Janzen & Applebaum (1979) indicated the avoidance of plant toxins as a major problem for animals that eat a wide range of foods in nature. Negative hedonic reactions to foods with bitter tastes is one means to avoid this problem. The other mechanism is through learning, a topic that will be discussed in a later section.

ANALYSIS OF THE INNATE HUNGER/APPETITE FOR SALT

When early forms of life emerged from the sea, the organisms lost a continually accessible source of sodium. Thus, land-based animals faced selection pressures that resulted in physiological mechanisms to conserve sodium and behavioural mechanisms for seeking and identifying sodium in the environment. Sodium is essential for the regulation of extracellular fluid balance and other basic functions, and because the body has limited sodium stores (Denton, 1982). For these reasons, it is not surprising that naturalistic observations and laboratory studies have revealed a strong preference for salty tastes among sodium-deficient animals. Early work by Richter (1943) demonstrated in rats that the removal of the adrenal cortex gland resulted in their deaths because of the sodium depletion. This gland secretes the hormone, aldosterone, which causes the renal system of the kidneys to resorb water and thus enable the body fluids to conserve the remaining sodium. However, adrenalectomised rats can survive indefinitely if given access to sodium salts which they drink with great avidity. Richter (1936, 1956) suggested that this specific appetite was innate, and offered the following ideas about its mechanism. Sodium deficiency induces changes both in sodium receptors and the central processing and evaluation of input from taste neurons signalling the presence of sodium. He hypothesised that a state detector sensed sodium deficiency and influences the preference for innate sodium. Subsequent research documented physiological changes in the gustatory system resulting from sodium deficiency, namely a lowering of the preference threshold for sodium (Contreras, 1977; Jacobs, Marks & Scott, 1988).

Salt hunger, or the appetite for salt (NaCl), is the model neurobehavioural drive system for illustrating innate specific hunger (Denton,

1982; Richter, 1936, 1956; Wolf, 1969). Rats show enhanced ingestion of NaCl the first time they experience sodium deficiency (Epstein & Stellar, 1955; Nachman, 1962) and respond within seconds upon first exposure to NaCl (Nachman, 1962; Wolf, 1969). In addition, sodium-deficient rats show the same immediate response to lithium salt (LiCl), which tastes almost like NaCl but which does not reduce sodium depletion (Nachman, 1963; Schulkin, 1982). All of the aforementioned observations argue against the notion that sodium preference is based on the after-effects of sodium ingestion.

There is a distinct hedonic component to NaCl appetite. If a rat is depleted of body sodium, it shows a change in its preference of liquids. It normally shows an aversion to drinking saline solutions which have concentrations greater than that of the body fluids (0.9% NaCl). For example, when offered a choice between water or hypertonic saline (i.e. greater than 0.9% NaCl) it prefers water to the saline solution. However, if its body sodium is depleted as a result of adrenalectomy or diuretic drugs (that cause excessive urination with the consequent lowering of body salts including sodium), the rat will show a behavioural preference for the hypertonic solution (Carr, 1952). There is also a more direct method of demonstrating hedonic changes arising from the ingestion of NaCl. Berridge *et al.* (1984) used the oral–facial expression method to study changes in the rat's affective reactions to various chemicals. They found that when rats were rendered sodium-deficient, their reaction to intra-orally infused concentrated saline is changed from a 'mixed oral–facial acceptance–rejection profile' shown under normal state to a positive facial display. This change is evident upon their first exposure to NaCl after their first bout of sodium deficiency. Sodium-depleted humans also report that salty foods are more pleasing (Beauchamp *et al.*, 1990). Such results indicate that the hedonic response appears to be unlearned and is part of the innate sodium hunger.

There is evidence for sodium hunger in humans. An early case study by Wilkins & Richter (1940) documented details on a young boy with adrenocortical insufficiency who showed a strong NaCl preference and a preoccupation of things salty as early as one year. He was brought into the hospital when 18 months old for observation because his parents were concerned with his abnormal craving for foods such as pickles, pretzels, salted crackers, potato crisps, olives, crisp bacon. Before he was able to talk, he would point to a cupboard containing a salt shaker. Because of his adrenal dysfunction, aldosterone was not produced, and he lacked the internal mechanism to regulate his sodium level or balance. This led to the implementation of the only available means, the regulation of sodium level through a behavioural mechanism and psychological preference. However, he was not required to learn that NaCl was necessary for his survival; he just showed strong positive reactions to its taste in a manner similar to that of the adrenalectomised rats. Booth (1994, p. 28) argued to the contrary, and suggested that the boy's craving for NaCl 'could have been a habit he had developed by stumbling on a way of reducing the malaise of adrenocortical insufficiency'. The unfortunate part of this case study is that when taken into the hospital, the boy was not allowed to eat NaCl, and died. Autopsy revealed that he had bilateral tumours in his adrenal glands which prevented him from secreting aldosterone. Detailed knowledge about aldosterone and the adrenal cortex was not available to the hospital staff during that early era of medical science.

Although sodium hunger may be mediated through innate factors, this does not mean that learning does not play a role in sodium identification. Classic studies by Krieckhaus & Wolf (1971) and Krieckhaus (1970) found that thirsty rats can learn the location of sodium sources, how to acquire sodium, and with what it is associated. When subsequently sodium deprived, the rats are able to display this knowledge. It should be emphasised that such results also indicate that rats are prepared to recognise the significance of salt. It is part of their innate endowment. In the Krieckhaus and Wolf experiment, thirsty rats were trained to bar press for dilute sodium salts, or water. When sodium-depleted and no longer thirsty, the rats previously exposed to saline demonstrated greater bar-pressing under extinction (non-reward) conditions than those who had been exposed to the water. This happens despite the fact that neither solutions were available in the test, an effect replicated by Dickinson (1986) and Weisinger, Woods & Skorupski (1970).

Similar effects were found in a study using a different method involving choice behaviour in a T-maze (Krieckhaus, 1970). Rats that had never experienced sodium deficiency were trained while water deprived to go to one side of a T-maze either for water or for a hypertonic saline solution. Then half of the animals were depleted of sodium and the control animals were injected with isotonic saline. During the following day the rats were tested in the apparatus, but responses to either side were not rewarded. Thus, this experiment involves a situation where the experimental group was not depleted of sodium during the training period and where sodium was *not* present during the test period. Under these conditions, these rats did not have the opportunity to associate NaCl intake with a sodium-depleted state, and thus did not have the chance to learn that NaCl would be beneficial. The fact that the experimental group showed prolonged running to the place where NaCl had been available (and the control group did not) suggest that the former exhibited an innate recognition of sodium.

Sodium appetite or sodium hunger?

An animal's intake of sodium can be increased without apparent depletion of this substance. Early studies by Fregly & Waters (1966) and Wolf (1965) have indicated that an increase in mineralcorticoid hormones from the adrenal cortex can stimulate sodium intake independently of body sodium homeostasis. Normally, the hormone, aldosterone, is released when the organism is deficient in body sodium because of its sodium-conserving properties. Paradoxically, Wolf & Handal (1966) found that subcutaneous injections of aldosterone into rats can produce an increase in saline intake. The fact that the aldosterone-injected rats were not depleted of sodium indicates that their ingestion of saline was not caused by 'hunger' but by 'appetite'. The authors suggest that the concept of 'sodium appetite' is more appropriate and accurate in accounting for their finding. Consequently, this concept has been used as a generic term in describing the 'motivation to seek, obtain, and consume salty tasting fluids and foods' (e.g. Stricker & Verbalis, 1990, p. 387).

Subsequent work by Weisinger & Woods (1971) suggested a mechanism underlying the aldosterone-induced 'sodium appetite'. They administered various doses of aldosterone and assessed their effects on the rats' intake of hypertonic saline. Both plasma sodium and plasma volume were increased following administration of the hormone. Rats raised with saline solution continuously accessible in their cages from birth did not show sodium appetite following administration of aldosterone unless some minimal time had elapsed wherein they were denied access to sodium. These data, along with those of control experiments, suggest that aldosterone-induced sodium appetite is the result of some associative process. Among the saline-reared rats, a state of low body sodium would never occur, and they had the opportunity to associate increased aldosterone with sodium appetite. In contrast, the control water-reared-rats would occasionally experience low or decreasing body sodium and elevated aldosterone release. This would induce sodium appetite, and, after several occurrences of low sodium, an association would be formed such that the presence of increased aldosterone would lead to sodium appetite in the absence of any sodium deficiency. Thus, the aldosterone assumes the properties of a conditioned stimulus.

The nervous system and the expression of sodium hunger/appetite

Schulkin (1989) posited that the lateral hypothalamic (LH) region regulates sodium motivation or drive (the 'internal state detector'), and that the central gustatory system at the level of the thalamus helps guide ingestion through the identification of NaCl (the 'food recognition detector'). Large lesions in either area abolish sodium appetite (Wolf, 1967, 1968; Wolf & Schulkin, 1980). However, rats simply exposed pre-operatively to the taste of NaCl were protected against the deficits in sodium appetite that usually results from lesions in the thalamic gustatory region (Ahern, Landin & Wolf, 1978). Merely sampling NaCl pre-operatively was enough to protect them. Paulus, Eng & Schulkin (1984) explain the reason for this protective effect. In the Ahern *et al.* (1978) study, the NaCl solution was located in the same place both pre- and post-operatively. The rats may have been protected because they remembered tasting NaCl in a particular place. Recall the results of the Krieckhaus (1970) and Krieckhaus & Wolf (1971) studies. Despite damage to gustatory sensibility, the rats returned to the place, guided by the memory of NaCl and its location in space. Paulus *et al.* (1984) tested this hypothesis by changing the location of the NaCl solution post-operatively for one group of rats and keeping it the same for another group. Their experiment clearly showed that only those rats with the NaCl solution in the same place pre- and post-operatively were protected. Rats with central gustatory damage remembered where the NaCl was located.

The effects of pre-operative NaCl drive arousal preceding lesions of the LH was studied by Ruger & Schulkin (1980), who demonstrated that LH-lesioned rats with no history of drive arousal do not exhibit sodium appetite while those in whom the drive was aroused pre-operatively do. A subsequent experiment by Wolf, Schulkin & Fluharty (1983) indicates that pre-operative NaCl drive alone without ingestion also protects the lesioned rats. Schulkin (1989) proposed that the amygdala may be critical for the protective effects of both LH and thalamic damage. Neither central gustatory nor the LH region are necessary for the expression of this behaviour, once the appropriate information is conveyed (tasting NaCl and the drive for NaCl) to the amygdala. We had encountered this structure in Chapter 3 when studying Fleming's (1989) analysis of maternal behaviour. Olfactory–gustatory inputs project to the central nucleus of the amygdala and facilitate the animal's evaluation of the significance of the stimuli contributing to these inputs. The central gustatory lesion at the level of the thalamic nucleus disconnects afferents en route to the amygdala, and impairs the significance of the taste of NaCl. However, if the thalamic-lesioned rat tastes NaCl pre-operatively, the taste pathway is no longer necessary. The rat recalls where NaCl was located and returns there.

Physiological basis of sodium appetite

Stricker & Verbalis (1990) developed a model of sodium appetite that shares features in common with general models of hunger and thirst. Ingestive behaviour is controlled by multiple excitatory and inhibitory stimuli. There is a distinct mode of central nervous system activity in which sodium appetite is facilitated. This mode is associated with sodium deficiency and/or hypovolaemia (severe loss of blood), and reflects a heightened sensitivity of the brain to signals of sodium appetite that would not elicit NaCl intake under conditions of sodium sufficiency. The view that brain function is influenced by availability of body sodium is conceptually similar to the notion of a set point for sodium homeostasis (Hollenberg, 1980).

Stricker & Verbalis (1990) also hypothesised that sodium appetite is controlled by both excitatory and inhibitory stimuli. The taste of NaCl is a permissive excitatory stimulus for sodium appetite. Like hunger and thirst, there are a variety of signals derived from changes in substrate levels, neural inputs to the brain and endocrine actions that provide specific excitatory or inhibitory stimulation of NaCl consumption. Just as thirst is stimulated by manifestations of water deficiency (details in Chapter 6), sodium ingestion is stimulated by manifestations of sodium deficiency including hypovolaemia, osmotic dilution of body cells and elevated levels of the hormone, angiotensin. The latter substance is released as a result of hypovolaemia and elevates blood pressure through pressor effects on the blood vessels.

In retrospect

Although we all need sodium for our organs to function, most of us do not have to worry about consuming enough sodium (unless we have adrenocortical insufficiency). Typically, modern diets contain much more sodium than we really need, and this reflects our innate liking of its taste. Fast-food restaurants exploit this human tendency, and many of us fall victim to their products much to our detriment. Consequently, the excessive intake and liking of NaCl results in a high incidence of hypertension (high blood pressure) amongst modern humans, which if left untreated, is associated with cardiovascular problems such as stroke (the breaking of a blood vessel in the brain). Hypertension develops from the effects of systems in the body to deal with the effects of ingested sodium on osmotic pressure and the volume of blood in our circulation. Instead of trying to retain it, our bodies are challenged by the excretion of excess sodium.

The fact that salty tastes have such strong attraction for us indicates how important this substance was in the diet of our ancestors. Before humans discovered methods of locating and producing NaCl in large quantitities, so that it is now very inexpensive and easily accessible, NaCl was a very valuable and treasured commodity. The strong influence of proximate mechanisms responsible for our tendency to like and ingest more NaCl than is 'good' for us is another example of the influence of the selection of past adaptations. Tooby & Cosmides (1990a, b) argued that adaptations to selection pressures of the past do not always result in adaptive behaviour of humans under current conditions. If the current environment differs from that existing during which the selection of proximate mechanisms for these problems occurred, there may not be a good match between these mechanisms and adaptive behaviour.

ANALYSIS OF THIAMINE-SPECIFIC HUNGER: ACQUIRED PREFERENCES

Thiamine-specific hunger refers to the enhanced ingestion of thiamine or vitamin B_1 by thiamine-deficient animals. The first experimental demonstration of vitamin B deficiencies and consequent selective behaviour was developed by Harris *et al.*, (1933). They found that when given a choice of three foods, including one supplemented with B vitamins, B-deficient rats would eventually choose the 'correct' food. Later, when the various vitamins of the B-complex were discovered and synthesised, researchers were able to create a specific thiamine deficiency. This permitted more precise analysis of thiamine-motivated behaviour. Thiamine is an important co-enzyme nececessary for the normal metabolism of carbohydrates and proteins. Deficiency of thiamine from the rat's diet results in diminished food intake with consequent body weight loss. In later stages there are also inflammation of nerves and motor impairment, but the signs and symptoms can be rapidly reversed through the administration of thiamine.

The basic experimental design for testing the effects of thiamine deficiency consists of raising weanling rats on a B_1-deficient diet for 21 days, and then offering them a choice between the deficient diet and the same diet supplemented with thiamine. Thiamine-deficient rats strongly prefer the thiamine-enriched choice, and control animals who are not deficient do not. Given such results, how could one be certain that the ability to seek and ingest needed thiamine is not innate? Rats made deficient in thiamine and offered a multiple choice of diets are incapable of making a correct choice immediately. We have noted in the previous section that sodium-deficient rats would respond immediately to foods containing sodium. In addition, Rodgers (1967) and Rodgers & Rozin (1967) succeeded in demonstrating that selection of thiamine-rich foods by a thiamine-deficient rat is not caused by specific innate factors. When B_1-deficient rats are offered the choice of a deficient but novel diet or the same diet supplemented with thiamine, they do not choose the latter, as one might expect. Rogers proposed that the salient factor affecting their behaviour was novelty of the diet and not its contents. Thus, in experimental tests that pit a new food against a familiar one associated with the deficiency, the new food preference may reflect attraction to the novel and aversion to the old diet.

Rodgers & Rozin (1967) found that thiamine-deficient rats prefer any new food over the familiar B_1-deficient one. The preference for novelty might account for the apparent immediacy of thiamine preference in deficient rats observed in most studies. The vitamin-enriched food also was the novel food. Rozin (1976) also suggested that the deficient rat's behaviour to the deficient diet when it is the only available food is similar to the behaviour of rats to innately unpalatable food. Furthermore, rats that have recovered from thiamine deficiency on a vitamin-rich diet refuse to eat the deficient diet when it is offered. Rozin uses these findings to explain the preference for thiamine-rich foods by recovered rats as a retained aversion to the deficient diet. A major feature of thiamine-specific hunger is learning that the deficient diet is undesirable.

Thiamine-specific hunger cannot be explained entirely as an aversion (Rozin & Schulkin, 1990). Thiamine-deficient rats can learn to favour thiamine-rich diets (Garcia *et al.*, 1967). This occurs because rats can associate an aftertaste with the beneficial effects of the source of the taste. A basic problem in learning about the properties of food is discovering which food or object or event is to be associated with either positive or aversive events. The modalities that are particularly associated with food identification, such as taste and smell, are primed to be associated with the consequences of ingestion (Rozin & Kalat, 1971). This is referred to as the 'belongingness principle' (Garcia & Koelling, 1966). The 'novelty principle' narrows the field to new foods. Another principle involves the idea of adaptive sampling that allows the rat to isolate the effects of individual foods (Rozin, 1969). All rats, whether deficient, recovered, previously poisoned or naive, tend to eat one food at a time. Although rats may sample multiple new and familiar foods when first exposed to them, subsequent meals tend to be primarily of one particular food source (Beck, Hitchcock & Galef, 1988; Rozin, 1969). Prior poisoning or deficiency exaggerates this sampling behaviour. Rzoska (1953) found that when rats sample a new food, they tend to take only a small amount, and that food neophobia is particularly evident in wild rats and is enhanced following poisoning experiences (see also Barnett, 1956; Richter, 1953) The sampling strategy provides the rats with the opportunity to test for aversive effects while avoiding lethal consequences.

All of the factors enumerated in the preceding discussion were documented in an experiment by Rozin (1969). He pointed out that the two-diet choice test used in most experimental studies is limited, well-defined and unnatural with respect to the situation faced by rats in the environment outside the laboratory. Rats normally live in a much larger area, have an elaborate social life and seldom face binary food choices. For this reason, Rozin analysed the meal patterns of deficient rats faced with the choice of a number of new foods. A thiamine supplement was placed in one of four diet choices for each rat. Feeding was restricted to eight hours a day, and the food intake from each cup was measured at hourly intervals. A characteristic pattern of feeding was revealed in the hourly and daily analyses. Meals tended to be restricted to one food, a pattern that maximises the possibility of associating each diet with consequences. This occurs because the meals tend to be isolated in time and to consist of a single food. Normal non-deprived rats show less well-defined sampling patterns.

Effects of other specific deficits

The major issues and explanation of thiamine hunger were definitively resolved in Rozin & Kalat's (1971) major review of the subject. Unlike most other issues in psychological research, most of the major fundamental questions and answers concerning the mechanism of thiamine hunger has been settled, due in large part to the creative energies of Paul Rozin. This led him to devote much of his subsequent research on the influence of culture and cuisine on human food selection, the topic of the last section in this chapter. Up until the present, we have encountered two distinct solutions to food selection problems in the rat – an innate NaCl hunger and a general-purpose learning mechanism for thiamine hunger. What about the other nutrients? Thirst, or the need for water, is like NaCl hunger. This is logical since water and sodium are basic and interrelated components of extracellular fluid. There is evidence for an innate internal state detector for water. The first time that chickens are water deprived, they drink an amount appropriate to their water deficit (Stricker & Sterritt, 1967). There is also evidence for a water detection system which is tied to the internal detector (Bartoshuk, 1968). However, there is no evidence of any other well-established innate system in rats.

Because the sensory quality of saltiness is correlated with the occurrence of mineral salts other than sodium, a variety of mineral salt deficiencies lead to enhanced preference for sodium (Schulkin, 1986). Salt licks usually contain several minerals in addition to sodium (Jones & Hanson, 1985). The innate hunger for sodium may serve as a guide for food selection in all mineral deficiencies. Although specific hungers for minerals such as potassium, calcium, iron, phosphorus, and zinc may be based in part on an innate bias to ingest salty tasting substances, general learning mechanisms guide subsequent reactions. Although rats tend to initially consume NaCl when potassium deficient, in the absence of a sodium source, they develop a preference for potassium salts (Adam & Dawborn, 1972). Indisputable proof that potassium-deprived rats do not show an innate preference for the needed substance has been revealed in the results of experiments by Adam (1973). In contrast to non-deficient rats, potassium-depleted animals show an aversion to the diet associated with the depletion and a strong preference for a novel diet. Their choice was not influenced by the presence of potassium in the familiar diet. In this respect, potassium depletion produced a novel diet preference similar to the effects of thiamine deficiency.

Specific hunger for all nutrients other than sodium may be mediated by the general-purpose learning mechanisms involved in thiamine hunger (Rozin & Kalat, 1971; Rozin & Schulkin, 1990). The marker for the operation of such a system is the presence of anorexia in deficient animals, a sign of learned aversion. When considering the essential need for protein and its low level in plant foods, we might wonder whether learning mechanisms mediate selective ingestive behaviour in deficient animals. Although Deutsch, Moore & Heinrichs (1989) demonstrated an immediate taste-mediated protein preference in protein-deprived rats, most of the other evidence favours a learning model. Rats are able to avoid amino acid imbalanced diets and to select diets containing limiting essential amino acids (Booth & Simson, 1971; Harper, 1967). This occurs because rats learn to like a taste, smell or texture in the presence of an incipient amino acid deficit when those sensory and metabolic cues have been paired with the correction of that deficit (Baker & Booth, 1989). Omission of protein from a single meal is sufficient to set up both the motivating bodily cue or depletion and the reinforcing effect of repletion.

Food selection under need-free conditions

Can animals and humans self-select a balanced or optimal diet in the absence of any deficiency and without any prior experience? The answer to this question comes from two sources. The early studies by Davis (1928, 1939) involved weaned human infants who were offered the choice of a variety of wholesome foods during a period of time (months). Infants indicated their preference by pointing at particular foods in the display of 15, and the nurse offered a spoon of the desired food. Davis reported that children ate a balanced diet, were healthy and grew normally on this schedule. It should be noted that all of the foods used in the experiment were of high nutritional value. It is possible that chance selection of foods such as dairy, meat, vegetable and fruit products would facilitate growth. Also, flavourings such as sugar were not added, and Rozin (1976) remarked that the most favoured foods were milk and fruit, the two sweetest items among the choices. These considerations raise doubts about the capacity of infants to self-select needed foods unless they are provided with an assortment of nutritious, equipalatable foods.

Animal studies using the 'cafeteria method' have produced results that question whether definitive adaptive food selection occurs under a need-free state. Lat (1967) reviewed the literature and found many examples of failure to regulate. Because palatability is a factor in food choice, the experimenters'choice of foods or nutrients in the test could have a strong influence on the results. In addition, Galef & Beck (1990) pointed out that need-free self-selection appears to be adaptive mainly in benign environments. Even in cases where there is adequate self-selection, incipient deficiencies may have been experienced which engage a set of innate or learned mechanisms. Thus, Rozin & Schulkin (1990) proposed that deficit-correction mechanisms may be responsible for the results.

Although this proposal can explain Davis's (1928, 1939) results, and is consistent with the outcome of Rozin & Kalat's (1971) analysis of food sampling strategies in thiamine-deficient rats, animals may not necessarily experience incipient deficits during the test. Research on feeding mechanisms discussed in Chapter 4 indicate that animals often eat, not because of a state of depletion, but because of cues in the situation that evoke ingestion. Under normal conditions, most ingestive behaviour is regulated by learned responses that have been acquired to defend homeostasis (Ramsay *et al.*, 1996). Similarly, anticipatory mechanisms may be responsible for the animals' ability to show appropriate food selection in the cafeteria test. This implies that although the animals may have experienced specific deficits and corrected them by selecting the needed nutrients, thereafter, the depletion/repletion mechanism is superceded by associative mechanisms.

CULTURE, BIOLOGY, AND HUMAN FOOD CHOICE

Aside from the influence of innate and learned factors, human food selection is shaped by the influence of the cultural setting under which a person is socialised (Rozin, 1976). For humans, the search and preparation of food and its ingestion occurs under a social setting. The earliest significant events in our lives include nursing and, later, weaning. Thus, from the very first, the taking of food is social. The cultural influence on food selection is transmitted through cuisine, a term that is comprised of two aspects. It can be used to refer to specific dishes and how they are prepared. These aspects include components such as the basic ingredients used (e.g. rice, potatoes, fish, meat), the characteristic flavours employed (e.g. the combination of garlic, olive oil and tomato for southern Italian, soy sauce, black bean and ginger for Cantonese Chinese, chili pepper and tomato for Mexican, curry for Indian) and particular modes of food preparation (e.g. fried, steamed, boiled, broiled). These three components describe the properties of the food served as the main course in most of the world's cuisines. In addition, there are rules about the order in which dishes appear within a meal, for instance, whether the salad is served before or after the main course, whether the soup is served before or along with the main course, etc., as well as what can be served with what, and what is is to be served at particular times or occasions. All of these features of food are culture-specific.

When studying the particular foods a person likes and eats, the researcher can learn a lot by asking that person one question: 'What is the culture or ethnic group of your family?' Rozin (1976) asserted that there is no other single question that would elicit an informative answer. The psychological mechanism mediating the strong influence of culture and cuisine on human food preferences is 'experience'. In rats (Galef, 1977) and cats (Wyrwicka, 1981) early food experiences are known to affect later food preferences. Galef suggested that the mechanism for these preferences may be familiarity, whilst Wyrwicka considered the role of parental factors in animal food preferences. Undoubtedly, both factors are influential along with socially acquired information from mother to children (Galef & Beck, 1990) on the development of food preferences. There are many edibles that we do not ingest unless they have been present in our mother's kitchen. For example, unless one had been raised in an Chinese household, it is highly unlikely that one would regard snakes, eels, sea urchin, sea cucumber, pig stomach and intestine, dried lichen or fungus as delicacies. Not only is 'mere exposure' (Zajonc, 1968) influencing the child's food preferences, but parents are active forces in teaching children about foods. However, Birch *et al.* (1982) found that parental attempts to shape children's food preferences are not always successful among American children. This conflicts with the food imprinting hypothesis, and may reflect the fact that children raised under Old World conditions are more susceptible to parental influence. In contrast, studies on American pre-school children have indicated that their food preferences are influenced by those of admired others, such as peers (Birch, 1980), or a nursery school teacher (Birch, Zimmerman & Hind, 1980). Such results reflect the influence of social forces on children's food preferences.

The 'biological wisdom' of cuisine

Cuisines and food attitudes have their own evolutionary histories, and particular cuisines may ultimately be accounted for in terms of the individuals who cumulatively give rise to cultures (Rozin, 1976, 1982, 1996). This approach explains socio-cultural rules about food in terms of biological features of the human omnivore. Rozin suggested that cultural information about foods is itself a distilled form of nutritional wisdom gained by individuals in the past by the learning process. There are two pathways through which biological aspects of humans influence cuisine. The behaviour path originates with biologically determined aspects of human food selection-behaviours that guide the evolution of cuisine. The metabolic path originates with biologically determined features of metabolism and nutritional needs. These pathways intertwine and result in behaviours or traditions that ensure health and minimise illness as well as weaken those that are detrimental. This explains the reasons for the constraints and guidance on food choice of cuisines.

The metabolic pathway to cuisine explains the physiological basis for the notion that cuisine represents adaptive choices. If there are substantial nutritional advantages to particular types of food processing, selection or combinations, humans will discover them. Rozin (1976, 1982, 1996) provided a detailed illustration of this proposition by analysing adaptive human practices in three different cultural or ethnic groups. A great many humans cannot digest a major component of cow's milk, lactose, after infancy (Simoons, 1970; 1978; 1982). Such lactose-intolerant people possess the ability to produce the enzyme lactase while nursing from the mother's milk, but lose the ability when it is deprogrammed during the period after weaning. This condition is common among natives of China, Africa and the Americas, whose consumption of raw milk leads to fermentation in the intestine caused by bacteria that break down the lactose. With the attendant discomfort experienced from gas, cramps and diarrhoea, such people learn to avoid drinking milk. This is particularly true in China, which is a lactose-intolerant culture where soya bean milk products are substitutes for milk from the cow. Other cultures have discovered a way to digest lactose externally, and thus allow for the ingestion of dairy products. If milk is simply left around for a few days at normal room temperature, bacterial action cleaves the lactose into two digestible components, glucose and galactose. This fermenting process led to the formation of cheese and yogurt. Yogurt is common in many eastern European cultures, and is an important staple in India.

Fortunately for most people of northern European origin, they have genes that block the deprogramming of lactase production during the weaning period (Simoons, 1982). Thus there is a prevalence of cow's milk-drinking among people from such stock. Although biological factors influence the culture and its cuisine, the inverse process can also occur. Later in human history, the domestication of animals made milk available as a food for non-infant humans. However, the usefulness of raw milk was limited by lactose intolerance in some. There may have been selection pressure; in cultures where animals that produce milk were abundantly available, there were mechanisms leading to the ability to digest milk. There may have been selection for genes that blocked the deprogramming of lactase at the time of weaning, with the result that some portions of the population became lactose-tolerant. Thus, there may have been two pathways to the acceptance of milk – changing the product before it entered the body, and selection of genes that make the body able to handle raw milk.

Although cuisine practices reflect the genetic/physiological characteristics of people in the culture, the 'rules' can become over-generalised to the extent that their biological origins are lost. As noted above dairy products that are cooked or fermented (e.g. cheese or yogurt) can be safely digested by lactose-intolerant people. Despite this fact, such products are rarely eaten, if at all, by most lactose-intolerant Chinese. However, in parts of the Far East, such as Hong Kong, where there have been western influences in many spheres of life including those of cuisine, cream sauces have appeared in dishes such as stirred fried crab or lobster. Such dishes are not part of the classical Chinese cuisine but are now popular among the Chinese in Hong Kong as well as those in North America.

Another example of adaptive human practice with foods concerns the preparation of a basic staple food in Brazil, manioc or cassava, which has been exported to Europe and Africa by the early settlers. Although this starchy root plant grows well in the tropics, and is easy to maintain, there is a problem with it as a food source. The principal form, bitter manioc, contains toxic levels of cyanide. Recall the discussion in an earlier section on the correlation between bitter-tasting foods and toxicity. The traditional processing of manioc before Columbus's arrival in the Americas eliminates the water-soluble cyanide by grinding the manioc and rinsing it many times with water. Although the indigenous people did not have the knowledge and technology of modern chemistry to analyse the contents of manioc, and discover the presence of cyanide, they undoubtedly noticed the effects of the consumption of improperly prepared manioc. The effects of rinsing it in water before cooking it are easily assessed, and this practice became established in the ritual of preparing manioc.

The third example of adaptive cuisine concerns the preparation of corn or maize in the New World. This plant is native to the Americas and was exported to Europe and the rest of world. Corn is a staple product in many traditional American cultures but is less popular in parts of Europe, particularly France. Eaten by itself, corn is not an adequate complete nutrient because it is low in vitamin B_6 (niacin), has an inadequate pattern of essential amino acids and is low in calcium. The traditional practice of the Mexicans in making a flat cake called tortilla out of ground corn solves these problems. The corn is soaked in a solution with mineral lime (from ash or powdered shells) which contains a high level of calcium hydroxide. Then it is ground and fashioned into flat cakes that are grilled (Katz, 1982). The heating of the corn meal in alkali makes niacin more available from the corn and also releases for digestion more of the essential amino acids which are in low levels in corn. In Mexican cuisine the tortilla is consumed with beans and chili pepper. By themselves corn and beans contain inadequate amounts of essential amino acids, but together, they make an adequate protein source. Given the unavailabity of meat except to the wealthy, one can appreciate the adaptive wisdom of such a cuisine to the peasants.

Neither the corn-bean combination nor the technology of tortilla preparation were imported to Europe with the corn. The consequence is that corn never became a staple diet in most of Europe, and farmers fed it to their animals. For this reason, although the intent was humanitarian, the shipment of corn by the United States as part of an aid package to the Europeans following the end of World War II was not appreciated. The French were particularly incensed by the receipt of what they perceived (culturally) as animal feed. In addition, the strain of corn that is grown in Europe is not the sweet, moist and edible 'corn on the cob' variety that is consumed with pleasure by most North Americans. This enhanced the misunderstanding between the cultures. Similar problems occur when the western world sends milk powder and other products to hungry people in parts of Africa and Asia where there is a high incidence of lactase deficiency, thus lactose-intolerance.

The anthropologist Marvin Harris (1987) analysed culinary practices in terms of nutritional value or the optimisation of diet. He explains the prohibition of beef-eating among Hindus of India in nutritional terms; the cow is more useful as a dairy source than as a meat source. This analysis was also applied to the explanation of the origin of traditions such as the dietary rules among people of the Middle East. Both Moslems and Jews of ancient times showed fear and loathing of pigs, and attribute their reactions to the animal's 'filthy habit' of eating dung. The twelfth-century physician Rabbi Moses Maimonides maintained that food prohibited by the Law is unwholesome, and much later, in 1859, the clinical association between trichinosis and undercooked pork was established. However, Harris argued that all domestic animals are potentially hazardous to human health. In North America, there have been recent concerns about the dangers of undercooked hamburger, and in Europe, there has been concern about the possibility of 'mad cow disease'. So why is there such a strong abomination of the pig? Noting the recurrence of pig aversions in several different Middle Eastern cultures, Harris suggests the ancient Israelite ban was a response to recurrent practical conditions rather than to a set of beliefs peculiar to a religion's notions about clean and unclean animals. The ancient Phoenicians, Egyptians and Babylonians were just as disturbed by pigs. Harris suggested that the early Middle Eastern preference for cattle, sheep and goats was based on the cost/benefit advantages of ruminants over other domestic animals as sources of milk, meat, traction and other services in hot arid climates. It represents an unassailably 'correct' ecological and economic decison embodying thousands of years of collective wisdom and practical experience.

Non-biological symbolic aspects of food

Bread baked without raising by yeast is a central part of Jewish families' tradition of remembering the group's escape from Egypt led by Moses. This practice is celebrated annually at which time it is orthodox to clear all yeast out of the home during the time of Passover. Booth (1994) views this as a purely symbolic act because its only basis seems to be the overt wish to relive the use of a longlasting food prepared for the journey into the desert. The ban on yeast is confined to the Passover celebration period. How the concept of subsisting on unleavened bread during the Passover is achieved depends upon the individual's upbringing and the custom of the household within a local synagogue. Family and synagogue leaders formulate an appropriate application of the symbolic principle.

For Christian communion, wine is the blood of Jesus Christ, as he stated when telling his disciples before his execution to eat bread and drink wine in memory of his self-sacrifice. Red wine is used in Christian traditions, and wafers of bread are made without yeast, because of the association of the Crucifixion with Passover. Christian denominations that avoid alcohol use unfermented red grape juice or other red-coloured juices such as blackcurrant. Not everybody is served both bread and wine in some churches. The key words in the Communion ceremony are the body that was broken and the blood that was shed. According to Booth (1994), the essence of the symbolism is a morsel to eat and/or a sip to drink.

This brief description of some aspects of Jewish and Christian religious practise with respect to food indicates that historical events are the basis of the symbolism and that biological factors play a minimal role in it. However, there are some metaphors with a physiological link, such as the Greek word for pleasure, *hedone*, relating to the Greek word for honey, *hedus*. The cognitive and sensory processses of experiencing pleasure from eating a sweet food is often used to describe people and/or music that we like.

SUMMARY

The ability of animals and humans to select appropriate nutrients and to avoid toxins in foods is mediated by mechanisms at various levels. For vital nutrients such as sodium and water, for which body levels are tightly regulated, there are nutrient-specific detectors in regions of the lateral hypothalmus that arouse appropriate food search, and taste recognition detectors of the gustatory–thalamic pathway that identify appropriate susbtances in the environment. There is also a hedonic component in which the 'liking' of salty tastes is increased when the organism's level of sodium is lowered.

For all other nutrients that are less tightly regulated, the selection of deficient substances is mediated by a multi-stage process involving malaise induced by the deficiency, the aversion of foods associated with the 'sickness', attraction of novel foods, and the preference for foods that are associated with recovery from the malaise. The extensive research on the behavioural regulation of deficiency of vitamin B_1 or thiamine provided such a model. Amongst humans, cultural influences shape cuisine in combination with their inherent likes and dislikes of specific tastes. In general most cuisines reflect adaptive solutions to specific biological problems faced by individuals in the culture.

6

Drinking activities

 Drinking is the means by which an organism acquires fluids necessary for the normal functioning of the cells in the body. Water is the medium through which the chemical processes of the body operates. It is the largest component of the body and its volume must be defended within narrow limits. 'The proportion of water to lean body mass (the body without fat) is essentially constant at 70%' (Rolls & Rolls, 1982). The energy processes of the cell occur within a fluid medium. In Chapter 4, it was noted that feeding preceded by food-seeking behaviour, is a necessity of life. Similarly, drinking preceded by water-seeking behaviour is also a necessity of life. Living organisms are endowed with mechanisms that have been selected for, and which cause them to be motivated to seek and ingest water when their internal environment's water balance is disturbed. The animal's nervous system is supplied with information from a sample of body fluids, and drinking decisions are based upon the state of this sample. Just as the thermostat samples temperature at a site in a room, it is assumed that the drinking mechanisms sample the fluid environment at one or more sites.

Fluid regulation in living organisms represents a balance between intake and excretion of water. Each side of the equation consists of a 'regulated' and an 'unregulated' component. The regulated component represents factors which act specifically to maintain body fluid homeostasis (water balance). The primary factors that regulate water balance are thirst and pituitary secretion of the anti-diuretic hormone (ADH), which is also known as vasopressin (Verbalis, 1990). This component of water intake reflects consumption of fluids in response to a perceived sensation of thirst. In contrast, the unregulated components of fluid intake consist of the intrinsic water content of ingested foods, the consumption of beverages primarily for reasons of palatability or secondary effects such as caffeine in coffee and the like. The unregulated component of water excretion occurs from water losses from sources such as evaporative respiratory losses in exhaled air, gastro-intestinal losses through diarrhoea and faecal water content, cutaneous losses in sweat as well as the amount of water that the kidneys must excrete in order to eliminate solutes generated by body metabolism. The regulated portion of water excretion consists of the renal excretion of free water above and beyond the amount necessary to excrete body solutes.

The water in the body can be conceptualised as being located in distinct compartments. There is the cellular compartment, known also as the intracellular compartment, which refers to the total quantity of water housed within all of the cells of the body. This water is located in tissues such as liver, heart and skin. All of the fluid outside of the cells constitute the extracellular compartment. A graphic representation of these compartments is presented in Fig. 6.1. The cell wall or cell membrane allows water to pass through with ease. Normally there are large flows of water in both directions across the cell membrane, and the cell volume remains constant because the flows are equal and opposite. However, if there is a flow in one direction, the cell must either shrink or swell. Osmosis refers to the movement of water across a semi-permeable membrane when there is a difference in water potential between the two sides. Water can pass freely through the membrane but it is difficult for a substance such as sodium to do so. A solution is said to be isotonic if it is of the same osmolarity as the fluid in the extracellular region. A solution of NaCl of about 0.9% by weight is isotonic with the plasma of the blood. If hypertonic (i.e. greater than isotonic) saline is injected into an organism, extra water is needed to restore the balance between solutes and solution in its body. The distribution of water in the intracellular and extracellular compartments under these conditions are shown in Fig. 6.1.

DRINKING AND THE PASSAGE OF WATER

When water enters the mouth, it passes along the oesophagus and into the stomach. Then it leaves the stomach, enters the intestine, and moves into the blood by the process of osmosis. If the fluid in the intestine is either pure water or hypotonic (i.e. less than isotonic) saline, water moves from the intestine into the blood. However, if the fluid is hypertonic, the osmotic pull is in the opposite direction – water moves from the blood plasma to the intestine. This explains why we experience intensive thirst if we drink sea water which is of a higher osmolarity than plasma. The balance between solutes and solution is disturbed, and

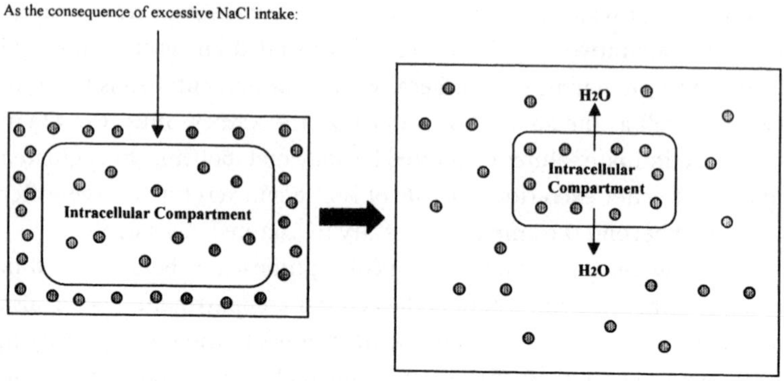

As the consequence of excessive NaCl intake:

Intracellular Compartment

H_2O

Intracellular Compartment

H_2O

Fig. 6.1 Distribution of body water in intracellular and extracellular compartments under normal water balance and as a consequence of excessive NaCl intake. (Figure courtesy of F. P. Valle.)

because of the rise in extracellular osmolarity, there is a compensatory movement of water from the cells.

Water is lost from the body through many routes. The kidney filters the plasma of the blood, removes waste products and excretes them as urine. In doing this, there is inevitably loss of water. The lungs are another source of water loss through evaporation. Water may also be lost as sweat, or in many animal species, through grooming and spreading saliva on the skin. The need to replace fluid lost from the body is regarded as the basis of the motivational mechanism of drinking. Through its assessment of information from the vascular system which provides a sample of the state of the body fluids, the nervous system generates drinking decisions.

A possible mode of feedback control is that through the mouth. A dry mouth sometimes causes humans to drink water. Grossman (1967) reviewed earlier literature indicating that the amount of saliva secreted by dogs decreases sharply during dehydration and correlates with body-fluid volume. However, one may question whether such a mechanism unaided by another detector can account for an animal's intake. The status of the dry mouth explanation of thirst is similar to that of the hunger contraction theory of hunger, which is discussed in Chapter 4. Both are found lacking. Even if the experimenter floods the mouth of the animal with water, the water-depleted animal will continue to drink. When the animal's oesophagus is surgically cut and moved outside of the body, the water drunk by the animal fails to change the state of the body-fluids even though the mouth is wet. Such animals show only limited and short-term satiety.

MECHANISMS OF OSMOTIC (CELLULAR DEHYDRATION) THIRST

A classic experiment by Gilman (1937) showed that cellular dehydration is a stimulus for thirst. He injected a solution of NaCl into the bloodstream of dogs. If the solution injected is hypertonic, as was Gilman's solution, then the extracellular fluid will have a higher concentration than the intracellular fluid. To dilute the concentrated fluid outside, water rushes out of the cells, which expands the extracellular fluid and leaves the cells dehydrated and shrunken. For this reason Gilman's dogs drank copiously. Similar results have subsequently been obtained in various other animals, including humans. Moreover, rats will run a maze in order to press a lever to obtain water after such injections. From such results, experimenters concluded that cellular dehydration arouses not just drinking *per se*, but a strong motivational state with its multiple specific responses.

The notion that prolonged water deprivation produces cellular dehydration has been established in past experiments (Adolph, 1939; Bellows, 1939; Fitzsimons, 1972; Gilman, 1937). Woods Rolls & Ramsay (1977) developed a technique which clearly demonstrates that there are receptors in the brain which respond to dehydration. They increased the tonicity of the blood vessels perfusing the brain, and by doing so, provided a more physiological stimulus than that arising from direct substances to the brain. The main blood supply to the forebrain is provided by the carotid arteries, and it is possible to infuse saline solutions into the artery. Woods *et al.* (1977) were successful in demonstrating a graded increase in drinking with increasing concentrations of saline in dogs. They demonstrated that there are receptors for thirst in the brain and that they are located within the area supplied with blood from the carotid arteries.

Osmolality is an expression of concentration reflected in a ratio of the total amount of solute dissolved in a given weight of water. Verney (1947) had also demonstrated, over 60 years ago that an increase in the osmolality of plasma elicits a prompt release of ADH which acts on the collecting ducts of the kidney to retain water. Studies by Robertson (1976) and colleagues indicated that normal human subjects in water balance showed an increase in ADH release following a two per cent decrease in body water and a suppression of ADH release when their body water was increased two per cent over baseline. This brings us to the question of how cellular dehydration may be sensed or metered so that ADH secretion, water intake and related physiological processes can be adjusted accordingly.

Osmoreceptors

Verney (1947) postulated that the effects of intravenous infusions of hypertonic saline solutions on ADH secretion may be mediated by osmoreceptors, which are excited by a decrease in their own volume or by the efflux of water across their cell membrane. These ideas have been extended as a result of the research by McKinley, Denton & Weisinger (1978) on mechanisms of cellular dehydration thirst. Johnson & Buggy (1978) and Thrasher, Keil & Ramsay (1982) located osmoreceptor cells in the vascular organ of the lamina terminalis (OVLT) and Verbalis, Hoffman & Sherman (1995) located others in the adjacent anterior hypothalamus near the third cerebral ventricle. Although the neural pathways connecting the OVLT with the ADH-secreting cells in the supraoptic region of the hypothalamus have been identified, Stricker & Verbalis (1999) suggested that 'the neural circuits in the forebrain that control thirst are still unknown'. Researchers during the past decades were more optimistic in suggesting the location of thirst-related osmoreceptor sites in the brain. In the rat, osmosensitive cells have been found in portions of the preoptic region, particularly the lateral area, now known to be directly involved in the control of ADH secretion. Subcutaneous injections of hypertonic saline modify the activity of spontaneously active cells in the lateral preoptic (LPO) regions. Nicolaïdis (1968) reported that some neurons in the preoptic area respond to hypertonic saline on the tongue as well as to intravenous injections. Also, some cells in the lateral hypothalamus of the rat are selectively activated by applications of sodium (Oomura *et al.*, 1969), and in the monkey, cells in this region decrease their firing rate when the animals drink water (Vincent, Arnauld & Bioulac, 1972).

Behavioural studies by Blass and colleagues (Blass, 1974; Blass and Epstein, 1971; Peck & Blass, 1975) also suggest osmosensitive thirst mechanisms in the LPO region of rats and rabbits. Micro-injections of hypertonic saline in sated animals elicit drinking while micro-injections of water in this region reduce the drinking reponse to systemic injections of hypertonic saline. Large lesions involving the LPO produce specific impairments in the animal's ability to drink in response to experimentally induced cellular dehydration (Blass & Epstein, 1971; Coburn & Stricker, 1978). However, Coburn & Stricker (1978) have argued that LPO is unlikely to be the only sensor for osmotic thirst. The effects of LPO lesions on water intake may be due to an interruption of fibres of passage, particularly the catecholaminergic components of the medial forebrain bundle. According to this view, LPO lesions reduce the normal drinking response to salt loads because the animal cannot adequately cope with the stress of the experimental treatment. For this reason, Coburn & Stricker (1978) argued that the function of the LPO is not specifically related to cellular thirst. More specifically, Stricker (1976) suggested that the induced disruption of behaviour was not specific to drinking but instead reflected a general inability of the lesioned animals to initiate movement. However, the study of McGowan, Brown & Grossman (1988) indicated that an interruption of fibres of passage may not be responsible for the effects of LPO lesions on water intake. They found that injection of the neurotoxin, kainic acid, into the LPO impaired the drinking response to cellular as well as to extracellular stimuli. This neurotoxin destroyed nerve cell bodies in the LPO without affecting the fibres of passage.

Summary of the role of osmotic stimuli and drinking

According to the cellular dehydration or osmotic thirst hypothesis, a critical stimulus for drinking is the loss of water from inside the cell. The receptors of this stimulus are called osmoreceptors, which are located in the brain. There are many sites of these osmoreceptors distributed along the preoptic and lateral hypothalamic areas, and it is likely that those for ADH release (in the supraoptic region and the OVLT) and those for the activation of water-seeking tendencies (in the LPO and lateral hypothalamic regions) are separate, though contiguous (Peck & Blass, 1975). Some of these brain regions are shown in Fig. 6.2.

MECHANISMS OF HYPOVOLAEMIC (OR EXTRACELLULAR DEHYDRATION) THIRST

Although there is no doubt that cellular dehydration is a factor promoting fluid intake, drinking can be influenced by other factors. The exchange of water between the body and the environment is actuated through the extracellular fluid. There are two components of this – intravascular fluid, which contains plasma, and the extravascular or interstitial fluid. Normally, these two phases of the extracellular compartment are in equilibrium and are maintained by the exchange of water and ions across the capillary walls. Most of the controls for maintaining fluid balance are located inside the vasculature. When plasma volume is low as a result of haemorrhage, stretch receptors and pressure receptors (baroreceptors) in the vascular system stimulate the release of ADH and thus renal conservation of water. Blood loss represents a loss of

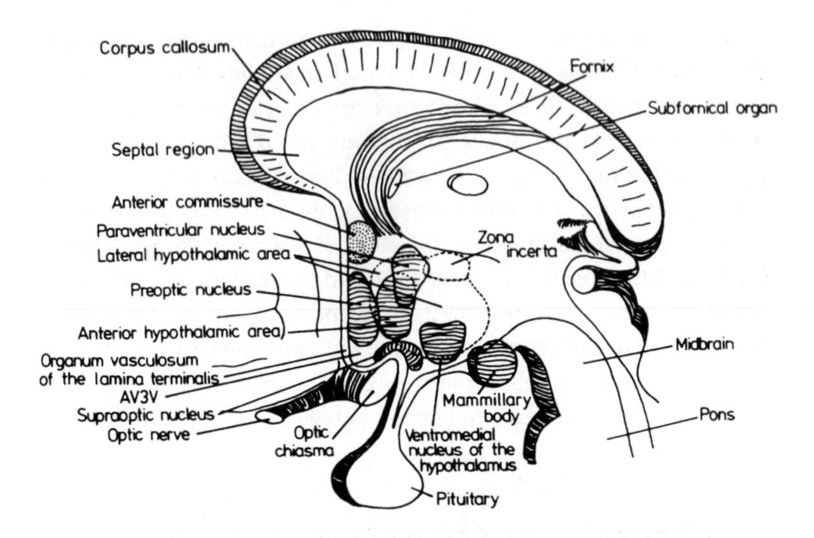

Fig. 6.2 Sagittal three-dimensional representation of the brain to illustrate some brain regions implicated in the control of drinking. AV3V: anteroventral region of the third ventricle. (From Fig. 7.1 in Rolls & Rolls, 1982, with permission from B. J. Rolls and E. T. Rolls, and Cambridge University Press.)

isotonic fluid which selectively depletes the extracellular fluid compartment with no osmotic effects. Fitzsimons (1961) found that haemorrhage in the rat stimulates drinking, following which Russell, Abdelaal & Mogenson (1975) discovered that water intake is related to the volume of blood lost. Such findings indicate that changes in the extracellular fluid compartment can stimulate drinking when there is no change in cellular volume. The critical depletion is reduced plasma volume or hypovolaemia. The final repletion of the extracellular compartment occurs when sodium appetite, arising from sodium loss in the blood, develops after the onset of thirst.

Angiotensin as a stimulus

Drinking, arising from extracellular depletion, is mediated by receptors in the vascular system. The decrease in blood volume is detected by baroreceptors and these signal other physiological and behavioural responses. Fitzsimons (1969) pointed out that all of the stimuli that produce extracellular hypotension also activate the renin–angiotensin system. Because of the reduction in blood pressure to the kidneys, the enzyme, renin, is released and acts on a substrate in the plasma to form angiotensin I, which is converted to angiotension II (ATII). ATII has many endogenous functions, the most important one being vascoconstriction, which serves to maintain blood pressure during hypovolaemia and hypotension. In addition to its hypertensive effects, ATII acts as a dipsogen, and as the findings of Fitzsimons & Simon's (1969) indicated intravenous infusion of it stimulates copious drinking. ATII elicits drinking by acting directly on specific receptors in the brain with this being the only behavioural response following its administration.

There is no dispute that ATII administered exogenously produces drinking in water-replete animals (Epstein, 1978; Fitzsimons, 1978). Under the conditions that normally produce ATII, there is, of course, a real body fluid deficit. Thus it is reasonable to assume that ATII and hypovolaemia detected by baroreceptors act synergistically to produce drinking. The behavioural effects of ATII are due to its actions in the brain. Because ATII does not penetrate the blood–brain barrier at normal pressures, the locations of these receptors must be in a region permeable to ATII. The blood–brain barrier refers to the fact that not all substances pass with ease from the blood into the brain. This barrier is not uniform throughout the nervous system, and in parts of the brain on the surface of the ventricles, there are receptors that can be reached by blood-borne agents. Several circumventricular organs have been suggested as receptor sites for ATII. Epstein (1978) and Simpson, Epstein & Camardo (1978) proposed that the subfornical organ (SFO) is one. Electrophysiological studies have identified ATII-sensitive cells in the SFO (Felix & Akert, 1974; Phillips & Felix, 1976). Local applications of ATII to the SFO stimulates drinking and the threshold dose is extremely low. Lesions of the SFO or local application of the angiotension inhibitor abolish drinking in response to intravenous ATII (Simpson & Routtenberg, 1973; Simpson, Epstein & Camardo, 1978). The systems regulating water and sodium balance as a result of hypovolaemia are depicted in Fig. 6.3.

There is evidence suggesting a role for both central neural as well as circumventricular sites that respond to ATII and thus enhance drinking. Swanson, Kucharczyk & Mogenson (1978) and Richardson & Mogenson (1981) proposed that both the preoptic area and the circumventricular organs contain receptors for angiotensin-induced drinking but which respond in different ways. Lesions in the lateral hypothalamus decrease drinking afer ATII injections in the preoptic area, but had little effect on drinking following ventricular injections. Similarly, Fitzsimons & Kucharczyk (1978) concluded, on the basis of experiments on dogs that there is angiotensin-sensitive tissue in both the SFO and the

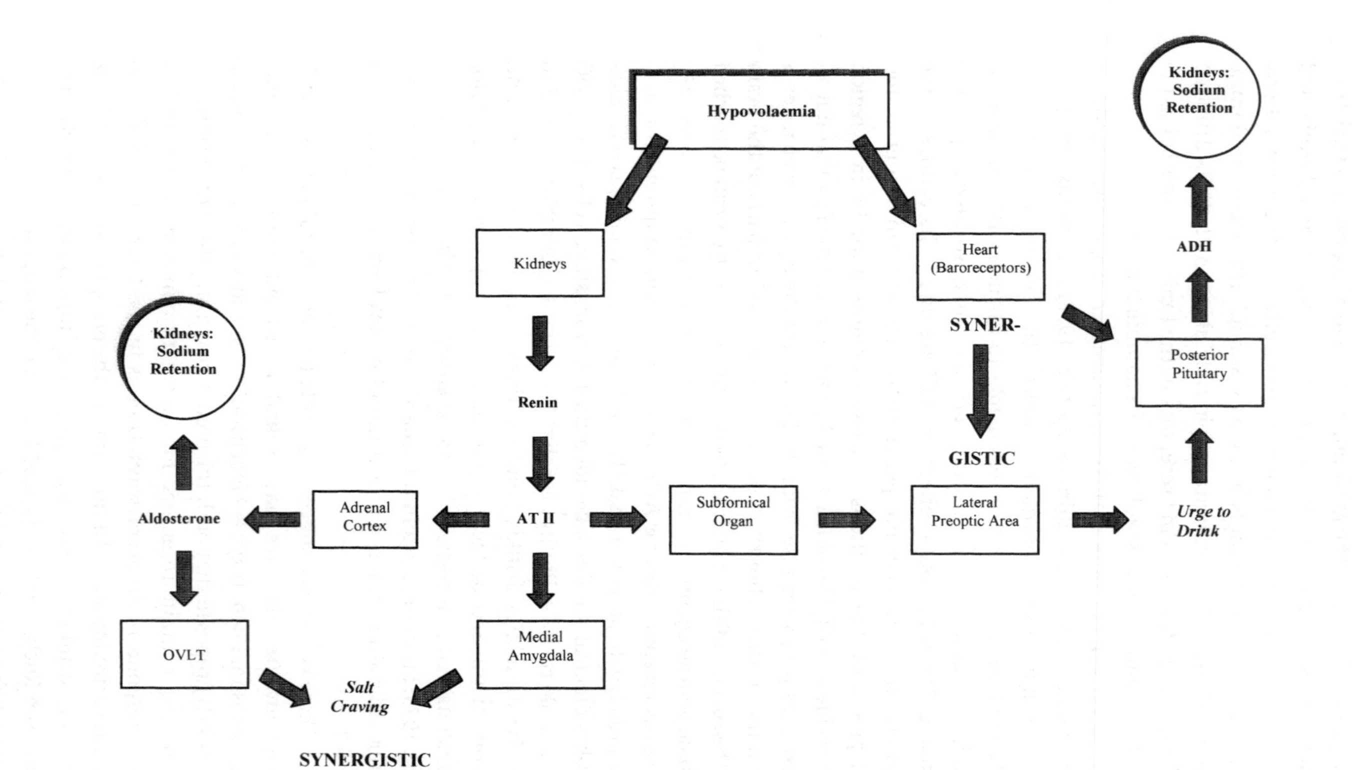

Fig. 6.3 The effects of hypovolaemia on systems regulating water and sodium balance. OVLT: vascular organ of the lamina terminalis. (Figure courtesy of F. P. Valle.)

preoptic area. Miselis (1981) found that the SFO sends efferent connections to another circumventricular organ, the OVLT and the medial preoptic area. These connections suggest that the three areas act as an integrated system for angiotensin-induced drinking.

Is angiotensin involved in deprivation-induced drinking?

Most experiments indicate that ATII is not an essential mediator of deprivation-induced water intake. The physiological actions of ATII can be blocked by administering substances which occupy the receptors normally occupied by ATII. The administration of the ATII blocker, saralasin acetate, does not affected water-deprivation-induced drinking in the dog (Ramsay & Reid, 1975), sheep (Abraham, Denton & Weisinger, 1976) or rat (Lee, Thrasher & Ramsay, 1981). Another approach is to lesion the SFO and observe whether drinking following water deprivation is affected. Although the lesion abolishes drinking in response to intavenous ATII, it had no effect on deprivation-induced water intake (Kucharczyk, Assaf & Mogenson, 1976). However, there is evidence that there is elevation of plasma ATII after water deprivation (Abdelaal, Mercer & Mogenson, 1976). In addition, Rolls *et al.* (1980) found that renin activity is elevated by water deprivation in humans and that it decreases to normal levels during rehydration. Such results make it possible that the renin–angiotensin system could provide an emergency mechanism for protecting plasma volume. Similarly, Stricker & Verbalis (1999) questioned whether ATII functions as a normal physiological response of thirst given that the doses of ATII required to stimulate significant water intake produce blood levels of ATII well above the physiological range.

NONREGULATORY DRINKING

The preceding discussion on drinking has considered it as a response to changes in the body fluids. Bolles (1979) pointed out that experiments employing traditional physiological techniques to study drinking results in models that are too simple to be relevant to 'real drinking behaviour'. The 'real rat' drinks for a variety of reasons. 'It drinks in connection with eating, and it drinks in anticipation of thirst. The rat in the cage will drink just because water is there. It may drink just because it has finished a meal or even because it has finished a bite of food (Falk's phenomenon). The real rat is a marvellously complicated drinker' (Bolles, 1979).

The issues raised by Bolles (1979) point out that drinking is not always in response to body fluid deficits, and that non-homeostatic factors are involved. He alluded to the close association between feeding and drinking. There are two possible reasons for this association – it is related to the actual physiological states and needs, and the animals may learn to associate particular smells and tastes of foods with particular fluid requirements, thus learning to drink in anticipation of fluid deficits. If animals are thirsty they tend to eat less. This may reflect the reciprocal relationship between the two states. Collier & Knarr (1966) examined the effects of a fixed daily water ration on body weight and food intake. The control group was given a daily ration of 36 ml water, typical of their *ad lib.* intake. Three other groups were given 22, 13 and 8 ml per day. After 80 days on these schedules, the groups showed marked differences in body weight and food intake wherein both dependent variables declined in proportion with their water ration. The amount of metabolically active tissue that can be supported by the animal is determined by its water intake, and is reduced when water intake is reduced. In essence, the animal is defending its water balance by reducing its food intake.

Because feeding and drinking are closely associated, 'spontaneous drinking' might be explained in terms of physiological changes associated with feeding. Fitzsimons (1972) found that rats with free access to food drank as much as 30 per cent of the water associated with a meal before the meal had begun. Thus drinking may be less a consequence of the physiological changes following feeding than the fact that the taste or smell of food or its physical presence in the mouth or stomach could cause drinking. Each food item is associated with a particular metabolic need for water. Animals may be able to associate the particular characteristics of a food with previously learned fluid requirements, and after several encounters, anticipate fluid needs. This hypothesis was questioned by Valle (1995) in the following manner. If water is freely available in such a situation, why is there any reason for the rat to avoid an impending deficit? It could drink later, when the deficit arises. Support of the secondary associative basis of normal drinking originated from Fitzsimons & Le Magnen's (1969) finding that rats drink more water while eating high-protein food, and their explanation that rats have learned that the ingestion of such food is followed, some time later, by a loss of body water. Because of the theoretical importance of this effect, Valle conducted experiments that were designed to probe, in greater detail, associative processes that were indicated in the diet-switch operation. In these studies he used a procedure in which the rats were given access to food for a limited time. Because it resulted in substantial daily food intake, he believed that such feeding habits would maximise the impact of feeding on body fluid loss and facilitate associations between the two consummatory acivities.

In general, Valle's (1995) experiments showed that under low meal-feeding conditions when rats were switched from a high to a low protein diet, water consumption dropped immediately to a value characteristic of low protein diets, a change that did not involve a learning process. When the rats were switched from a low to a high protein diet, their water intake during the meals increased immediately to a level characteristic of high protein meals. This effect was evident within the first 30 minutes of the rat's first encounter with the food. Again, there was no evidence that the rat's water intake during meals was influenced by a learned anticipation of subsequent changes in its water economy. Valle argued that although normal meal-related drinking in the rat does not reflect the influence of existing fluid deficit, the motivating force behind this secondary drinking may not involve anticipation of a subsequent fluid deficit. Instead, water intake during meals is controlled by the oropharyngeal properties of food. The rat's water intake is influenced by some hedonic factor that is related to its difficulty in chewing and swallowing particular foods. The greater the difficulty in processing the food, the greater the positive hedonic value of the accompanying drink. Valle suggested that meal-related drinking is primarily a hedonically driven option for the animal.

The influence of *hedonic variables* has been documented by a variety of methods. The addition of saccharin to the water of non-deprived rats produces increased intake which was not further affected by deprivation (Ernits & Corbit, 1973). Similarly, research by Rolls, Woods & Rolls (1980) indicates that other palatable flavours such as orange or cherry enhance the fluid intake of non-deprived rats. Conversely, the addition of an unpalatable substance such as quinine (see Chapter 5) reduces the fluid intake of rats (Nicolaïdis & Rowland, 1975). Another demonstration of the hedonic interpretation of normal drinking comes from experiments in which rats were given intravenous infusions of water. In terms of the integrity of the water balance, rats should show decreased intake of water. Rowland & Nicolaïdis (1976) found that despite the infusion of 180 ml of water a day, the rats persisted in 50 per cent of their normal water intake. Drinking was abolished only when water was infused during the rat's feeding and when the tube carrying water to the stomach provided oropharyngeal cues similar to those provided by the actual ingestion of water.

Recently, Kraly (1990) and Kraly *et al.* (1995) suggested a mechanism by which the occurrence of normal drinking can be linked to the presence of tangible chemical signals operating upon known neural mechanisms. His team demonstrated that water intake can be elicited by histamines and by naturally produced ATII with both effects occurring in the absence of fluid deficits. Histamine is a compound that serves as a transmitter substance in the brain and as a hormone-like messenger in the rest of the body. When an animal eats, cells in the stomach release histamine. An injection of this substance will cause drinking (Kraly, 1983), and because the secretion of renin is controlled by histamine receptors in the kidney, Kraly & Corneilson (1990) suggested that thirst produced by histamine may involve ATII. Kraly's (1990) model proposes that oral cues from food activate histaminergic mechanisms which in turn engender substantial intakes of water, before any change in extracellular fluid occurs. Exposure to cues associated with feeding activates the synthesis of ATII, which elicits a small amount of drinking before the meal begins. However, this model cannot, in its present state, explain the effects of high protein food on meal-related drinking. It may if there is some way to relate ATII levels to abrupt changes in normal drinking that occur when the type of food is changed.

Schedule-induced polydipsia

When a food-deprived rat is fed immediately after emitting a response, the food-producing behaviour is increased in strength. When water is concurrently available, hungry though not thirsty rats will drink even though the reinforcer is food. The phenomenon of hungry rats consuming copious amounts of water when fed small amounts on an intermittent feeding schedule is referred to as *schedule-induced polydipsia* (SIP) (Falk, 1961). In his initial experiments, Falk found that the rat's mean water intake during the three-hour sessions was over 90 ml, more than three times its normal daily fluid intake.

Certainly such drinking serves no water regulatory function. In fact such behaviour would serve to upset the water balance by causing severe overhydration. The previous discussion in this chapter suggests that drinking is in response to either relative or absolute dehydration. How is it possible that an animal in normal water balance shows such excessive drinking in Falk's situation? Can external environmental conditions such as an intermittent feeding schedule override the normal physiological modulation of consummatory behaviour?

SIP has been observed with a variety of procedures in which food pellet delivery is spaced over time. The effect occurs when a response is followed by food after some programmed time has elapsed following the delivery of the last pellet. Thus the spacing of the pellets is a critical determinant of drinking. In addition, large amounts of water are ingested if pellets are delivered independently of the rat's performance (Gilbert & Keehn, 1972). SIP does not depend upon an explicit relationship between a selected response and pellet delivery. Although SIP does not serve homeostasis, it does not normally upset fluid balance, given that polyuria accompanies polydipsia. It should be also noted that SIP occurs under very particular experimental conditions and it is unlikely to be a condition that animals will have encountered in their naural environment. Animals are unlikely to have evolved mechanisms for coping with this situation, and the situation does not necessarily relate to the controls of normal drinking.

A common denominator in all the SIP experiments is the necessity of food deprivation and the thwarting of eating. This combination of events can induce aversive effects in the rat. The thwarting of eating by spaced presentation of pellets results in excessive drinking, which is assumed to arise from the aversive effects of the infrequent occurrence of food. It has been proposed that frustrative effects and SIP are both reduced whenever the animal is freed from food deprivation. In SIP studies, the animals typically are held at 80 per cent of their normal, free-feeding body weights. This procedure has been followed in order to ensure that the strength of the feeding response remains high throughout the session. By allowing the body weight to rise, the experimenter can weaken the animal's feeding response. Falk (1969) accomplished this by gradually increasing the amount of post-session food supplements. He found that drinking remained high until the rats' body weight reached about 95 per cent of its free-feeding level. At this point, the animal's motivation level for food was low, resulting in the effects in which intermittent feeding was less aversive. A recent study by Lamas & Pellon (1997) produced similar results.

An alternative interpretation of SIP involves arousal level. Both food deprivation and intermittent food delivery induce arousal and it may be this arousal which leads to SIP. Support for this view comes from the finding that SIP reduced pituitary–adrenal activity which was elevated in situations of high arousal (Brett & Levine, 1979). The relationship between adrenocortical hormones and arousal is discussed in Chapter 6. In addition, adrenalectomised rats fail to develop SIP, but injections of corticosterone in these animals fully restore the acquisition and expression of this effect in them (Levine & Levine, 1989). Such results suggest that corticosterone plays an essential role in the normal acquisition and development of this behaviour. Similarly, Mittleman, Blaha & Phillips (1992) found that performance of established SIP was decreased by adrenalectomy, but that the effects of corticosterone on established SIP depended on the level of performance. High levels of drinking were enhanced by a high dose of corticosterone, whereas low rates of drinking were increased by a low dose.

Falk (1972, 1977) has suggested that SIP is equivalent to the displacement activity which occurs in more natural situations. SIP is but one of a number of activities that may occur under similar schedule conditions. These activities include wheel running, aggressive behaviour, chewing wood shavings and air licking. Falk considers these activities to consist of a class of behaviour that is adjunctive and related to the displacement activities studied by ethologists. Displacement activities are usually observed in situations in which the appropriate consummatory response cannot occur because of response thwarting. Staddon & Simmelhag (1971) also relate adjunctive and displacement behaviour. An animal working for food on an intermittent reinforcement schedule learns that during certain periods, such as the time immediately following food presentations, food is never available. At such times, animals tend to avoid the food-getting situation and its associated responses. The area of the test chamber does not allow the animal to leave the food-getting situation entirely, and thus a conflict situation results. If the required external stimuli are available in the situation, the consummatory behaviour appropriate to another motivation (such as drinking) may occur. For example, Levitsky & Collier (1968) found that rats that showed polydipsia on a variable interval food reinforcement schedule would run at a high level if given access to an activity wheel. When both water and the wheel were available, drinking occurred during pellet ingestion and running occurred throughout the interreinforcement interval.

The substitution of behaviours found during food interval schedules parallels Valenstein, Cox & Kakolewski's (1968) results with intracranial-stimulation-induced behaviours. They found that when eating, drinking or gnawing is produced by electrical stimulation of the hypothalamus, one of these behaviours can substitute for the one originally induced. This was accomplished by changing the options available to the animal while holding the hypothalamic stimulation parameters constant. We will re-visit this theme in Chapers 8 and 10. Such results together with those on feeding schedules indicate the extent of environmental control exercised by stimulus facilitation. To that extent there is merit in Falk's attempt to explain SIP in terms of the general class of adjunctive behaviour.

SUMMARY

Drinking enables the organism to maintain the water balance of the cells in the body. Osmoreceptors located along the supraoptic, preoptic and lateral hypothalamic areas of the brain detect the loss of water from inside the cells. Stimulation of the OVLT and supraoptic area trigger the release of the antidiuretic hormone which enables the kidney to conserve remaining body water. The lateral preoptic area and the lateral hypothalamus are involved in the activation of water-seeking tendencies. Drinking can also be induced by reduced blood plasma volume which stimulates receptors in the vascular system. The stimuli which produce hypovolaemia also activate the renin–angiotensin system, and fluid intake is enhanced.

Drinking is not always a response to body fluid deficits. There is a close association between feeding and drinking, and 'spontaneous drinking' may reflect the effects of the taste and smell of food or its presence in the mouth or stomach. Hedonic variables also influence drinking. Water that is flavoured with palatable substances such as saccharin, orange or cherry enhances fluid intake, whilst the addition of a bitter-tasting additive, such as quinine, reduces fluid intake. Drinking can also be manifested in hungry rats under normal water balance following the thwarting of an adjunctive response such as eating. The intermittent availability of very small portions of food elevates the animal's arousal level which is reduced by its high rate of drinking.

7

Stimulus seeking and exploratory activities

 Observations of animals placed into an unfamiliar environment indicate that they display a characteristic pattern of behaviour which suggests exploration. They typically move throughout the physical space and enter many parts of it. If unfamiliar objects are present in this environment, the animals may approach these objects and make physical contact with them. These actions are characteristic of stimulus seeking behaviour 'which serves to acquaint the animal with the topography of the surroundings included in the range' (Shillito, 1963). Such behaviour in which the animal familiarises itself with its environment may serve an adaptive role. By doing so, the animal acquires information that is potentially useful, such as discovering potential food sources and escape routes. The phenomenon is manifested in all mammals. The notion that familiarity of the environment assists solutions to problems encountered later is also relevant to humans using the World Wide Web. Users who surf the Web without any particular goal acquire information which may become useful to them in later contexts (Seltzer, 1998).

Russell (1983) suggested that there is more to be gained from immediate exploration of a new environment than from not exploring, and thus regards the former as an adaptive strategy. Not exploring would lay an animal open to the hazards of an unknown environment. In social animals such as the rat, exploration may also have the goal of establishing contact with conspecifics. This is often the case when the animal has been removed from a group for testing (Suarez & Gallup, 1985). Although there may be many reasons for exploratory activities, there is no doubt that information about the environment is being processed in the nervous system of animals whilst they are engaged in contact with it. Thus, Toates (1986) proposed that exploratory behaviour is provoked by the interaction of the state of the nervous system with the stimuli impinging upon its sense organs.

There are various proximate causal factors influencing stimulus seeking behaviour. An early theory by Fiske & Maddi (1961) proposed that both stimulus variation and stimulus intensity contribute to the total 'impact' of a stimulus. The amount of 'impact' sought by an animal depends upon its characteristic level of arousal. They hypothesised that the extent of stimulus seeking is dependent upon the animal's characteristic level of arousal. The latter can be manipulated by varying the amount of environmental stimulation that it receives during its infancy. In contrast, Berlyne (1960, 1963, 1967) proposed a different model of the effects of arousal and stimulus seeking, in which arousal is viewed as a state defined independently of stimulation. I will discuss research which has been influenced by these arousal models before presenting other explanations of the causes of stimulus seeking.

EARLY REARING CONDITIONS AND STIMULUS PREFERENCES OF MONKEYS

Experiments involving behavioural comparisons of rhesus monkeys that were reared under different conditions provide empirical support for the optimal stimulation and arousal level explanation of stimulus seeking. In his attempts to specify the nature of organism–environment interactions underlying the development of emotional attachment of these monkeys, Sackett (1965) studied the effects of rearing conditions on their reactions to novel stimuli. His three-year-old monkeys were reared either in total isolation from birth to one year, or alone in a wire cage for one year where they could see and hear other monkeys but not physically interact with them, or in the jungle and caught when about one year old. These animals were given access to an apparatus upon which they could climb but which varied in complexity.

The results of Sackett's (1965) study indicated that feral monkeys that had been reared with greater proprioceptive stimulation arising from the jungle conditions prefer stimuli with greater proprioceptive complexity. Isolate and wire-cage monkeys reared with less complex stimulation involving movement, preferred to interact with objects containing simple proprioceptive stimulation. In general, these results indicated that as the complexity of movement inherent in the response decreased, the isolate and wire-cage monkeys increased their number and duration of contacts of the apparatus. In contrast, the feral monkeys showed a decrease of exploratory activity when proprioceptive stimulation and complexity decreased. Singh (1969) extended this analysis by studying feral rhesus monkeys that lived under jungle (rural) and city (urban) environments in their native India.

Singh (1969) suggested that city and jungle habitats provide different opportunities for varied perceptual and motor experiences for these monkeys, and that conditions in the urban area are more stimulating than those in the jungle. This led him to perform experiments on monkeys caught in the bazaar areas in some Indian cities with those caught in the interior of the jungle. He tested these animals on an apparatus that presented them with three stimulus displays of different complexity levels. The results indicated that on this test for visual exploration, urban monkeys were more responsive to stimulus displays of higher complexity than those preferred by the jungle monkeys. Singh interpreted such results to indicate that urban monkeys operate at a higher complexity level than the jungle monkeys, and thus exhibited a greater degree of stimulus seeking to these stimuli. When taken together, the experiments by Sackett (1965) and Singh allow us to compare the effects of subnormal, normal and enriched environmental conditions on subsequent stimulus seeking behaviour. Each group seems to seek an 'optimal level' of arousal that varies depending upon the conditions under which they were reared.

EARLY EXPERIENCE AND STIMULUS SEEKING TENDENCIES IN RATS

Another set of experiments where the animals' characteristic level of arousal was manipulated through early stimulation are those involving rats. A series of experiments by DeNelsky & Denenberg (1967a,b) indicate that differences in stimulus seeking in adult rats are the result of infantile stimulation during an early period. In general, early stimulated rats investigate novel stimuli to a greater extent when adults than control rats who did not receive such treatment during the preweaning period. This effect reflects the influence of early stimulation on arousal level. Because of their lower level of arousal, the control rats show 'fear' or aversion during their initial contact with novel stimuli. In contrast, the 'treated' rats are attracted to such stimuli, which serve as positive incentive for sensory contact.

The treatment of infantile handling consisted of removing preweaned rat pups from their mother and nesting cage for short, daily sessions. Aside from showing a greater tendency to explore a novel environment, handled (H) rats were less emotionally reactive than nonhandled (NH) rats when placed in it. H rats also show acceleration of development of body hair, opening of the eyes, locomotion, body weight and puberty (Ader & Friedman, 1965; Altman, Das & Anderson, 1968; McMichael, 1961), and higher resistance to stress (Ader & Grota, 1969; Denenberg & Haltmeyer, 1967; Levine & Lewis, 1959). The early studies involved daily treatment during the 21-day preweaning period but later studies indicated that similar effects can be induced by stimulation over a period as short as the rat's three postnatal days (Levine, 1969).

Levine (1969) commented that manipulations such as the administration of electric shock at low intensities, and exposing the pup to cold stress were effective in producing a 'handling effect'. Levine provided a clinical sketch of H and NH rats in the following manner. When the experimenter walks into the laboratory, all of the H rats are in the front of their cage, and all of the NH ones are in the back. If the experimenter offers them a novel food such as a stick of celery, the rats will eat it within 15 minutes whereas the NH rats would not touch it during this period. If the rats are placed in a novel apparatus, the H rats will move back and forth whereas the NH ones will either stay in a corner or skip very fast and stay in a corner. This description indicates that there are dispositional differences between these groups, especially with respect to their reactions to novel stimuli and situations. The challenge is to assess whether these reactions can be separated through experimental analysis.

Following a battery of tests designed to assess dispositional difference, Whimbey & Denenberg (1967a,b) subjected the data to factor analysis, and extracted three meaningful behavioural factors. These were identified as: (1) *emotional reactivity*; (2) *field exploration*; and (3) *consumption–elimination*. The field exploration factor was defined by activity scores on a variety of measures including the open field. On the first day of the open field test, the NH rats were more active even though their greater level of defecation indicated that they were more emotionally reactive than the H rats. However, on subsequent daily test sessions, they were less active than H rats but continued to defecate more. Although emotional reactivity and field exploration were isolated as independent factors, there is an interaction indicating different manifestations, depending upon when the activity measure was taken. When their latency to emerge from a dark or familiar area into a bright or strange one was studied, H rats show accelerated emergence relative to NH rats (Hunt & Otis, 1955; Hunt, 1963; Levine, 1969). Such results indicate that early handling facilitates approach behaviour in novel settings (DeNelsky & Denenberg, 1967a, b). This effect indicates that H rats show faster habituation of fearful responses to novel environments. Denenberg (1969) has

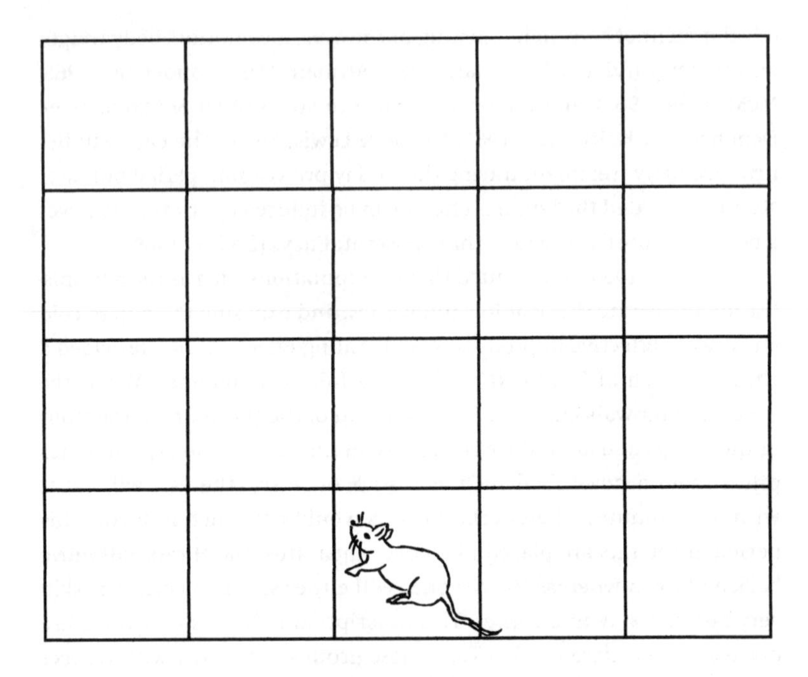

Fig. 7.1 The open-field test apparatus. (Drawn by Pouneh Hanjani.)

remarked that the use of a series of trials is standard practice because differences related to early stimulation often manifest themselves *only after* the first trial (also see Russell & Williams, 1973 on this issue). An open-field apparatus is shown in Fig. 7.1.

Exploratory or stimulus-seeking behaviour can also be measured in an operant chamber with the 'light contingent bar-press test' (LCBP). This is a method of assessing the rewarding effects of stimulus change. In this apparatus rats will press a lever in a darkened box in order to switch on a light of moderate intensity. A change in the stimulus conditions arising from the rat's action is assumed to reinforce bar-pressing behaviour. If the chamber is illuminated, rats will also press the lever to turn off the light, but not with as high a rate of responding as they would for the illumination of a dark chamber (Barnes & Kish, 1958; Glow, 1970). However, the level of illumination in such tests is an important factor. With extremely intense illumination, it would be more rewarding for the rat to terminate the stimulus than to initiate it in a dark box. Halliday (1967) suggested that light onset in a dark environment is rewarding partly because it facilitates visual exploration. It is accompanied by scanning head movemens. Even though change *per se* may be rewarding, Halliday (1967) suggested that light onset is more rewarding than light offset because it does not preclude visual exploration.

Wells *et al.* (1969) found handling differences when the rats were given the LCBP. They predicted that if stimulus variation declines with repeated exposures to a given environment, which also was suggested by Denenberg (1969), there should be exploratory differences between H and NH rats during the earlier but not later sessions. Their results showed that the H rats responded at a higher rate than NH ones during the first LCBP session, but on later sessions, the difference was not significant. This finding suggests that an understanding of the relationship between emotionality and exploratory behaviour might best be achieved by observing an animal's behaviour in response to stimulus change over time. This suggestion was followed up by Williams & Russell (1972) who tested H and NH rats for three minutes on each of ten consecutive days in an open field. They placed a small metal cylinder in the centre of the field, and measured the number of times the rats entered the second section as well as the total number of floor sections that they traversed. They found that H rats showed a straightforward increase in ambulatory activity over trials, whereas the NH rats were least active in the middle trials of the series. The experimenters interpreted this to mean that the first trial ambulations scores of the NH rats reflected escape behaviour, and those on later trials were motivated by their stimulus seeking tendency. This inference was derived from the observation that NH rats entered more peripheral sections on Trial 1 than on 2, 3, 4, and 5, but that they entered the centre section less on Trial 1 than on Trial 5. Because the experimenters interpreted entries into the centre as exploratory responses directed at an object, they hypothesised that exploration increased as the suppressive effect of 'fear' decreased following habituation to the metal cylinder. Both H and NH rats showed an increase in entries into the centre over the test sessions.

A methodological problem in using ambulation as well as response rate measures to study stimulus seeking behaviour is it is confounded with the effects of fear. Fearful animals are less able to express exploratory tendencies because they are more likely to freeze in a novel environment. A less ambiguous test of handling differences in seeking behaviour would be to use a choice rather than an ambulation or activity measure. This was done by Wong & Jelliffe (1972) who first tested H and NH rats in a T-maze where at the choice point, they could either turn in one direction leading to a small gray goal box containing food pellets or in the other direction leading to a large open field containing various novel objects. In this situation both the small goal box and the open field were novel to the rats, but there was more 'complexity' in the field because of the objects in it. During this phase of the experiment, the rats were tested under *ad lib.* feeding conditions. The results indicated that H rats showed a greater tendency to enter the open field than did NH rats, thus indicating a greater preference for complexity. However, when tested while food-deprived, the H rats immediately shifted their preference to the side leading to the small box containing food, while the NH rats required 30 trials before this happened. In the third phase of the experiment when the rats again had unlimited access to food prior to the test trials, the H rats immediately directed their responses to the arm leading to the open field. The NH rats again took about 30 trials before they shifted their choice to the arm leading to the open field. These results may be interpreted in terms of arousal level differences. The complexity of the open field served as a positive incentive for eliciting exploratory behaviour among the H rats because of their higher arousal level. The NH rats required repeated exposures to these stimuli before they were able to habituate to the novel and overarousing effects. It is interesting to note that when food-deprived, the H rats took less time to make an adaptive choice. The early Dember & Earl (1957) theory suggests that complexity is not a concrete attribute of the stimulus but rather a dimensionless property analogous to Berlyne's (1960) construct of *arousal potential.* Differences in arousal potential between these two groups may explain the outcome of the Wong & Jellife (1972) results. The apparatus used in this experiment is shown in Fig. 7.2.

Tests of spontaneous alternation behaviour (SAB), which were described by Tolman (1948), uncover the tendency of animals to choose the arm that has been least recently visited in the maze (Dember & Richman, 1989). This method provides another test of stimulus seeking which is not confounded by differences in activity level, and was used by Wong (1969) to study the effects of infant handling. With this method, the animals' reactions to novelty rather than to complexity are being assessed. The rats were given a free choice on Trial 1 of goal regions that were both black. On Trial 2, the arm on the previously unchosen side was blocked , which restricted (forced) them to enter the same side chosen on Trial 1. Between Trials 2 and 3 the black goal box of the previous non-entered side was changed to a white one, and on Trial 3, each rat was given a free choice. Odour cues were controlled in the apparatus by

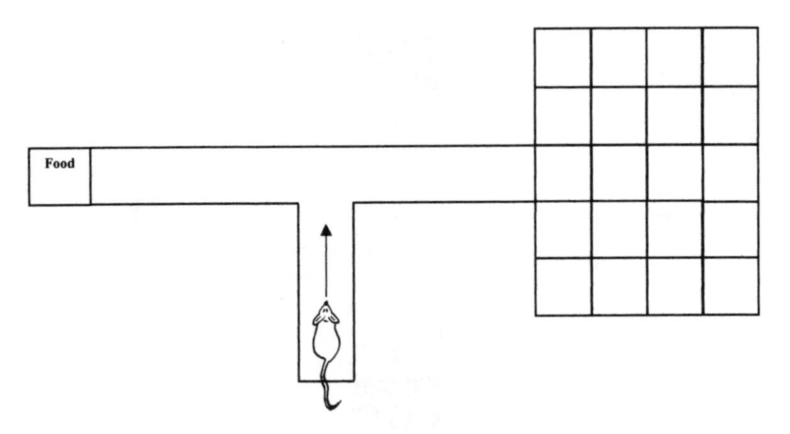

Fig. 7.2 Test of preference for food or sensory-activity incentives. (Based on Wong & Jelliffe, 1972.)

washing it with detergent and water following every trial. The results showed that when rats were tested under an *ad lib.* feeding schedule, both H and NH rats alternated at an above-chance level (79%). However when tested when food-deprived, the H rats alternated at chance level while the NH rats showed 72 per cent alternation. Hungry rats that are reared under 'normal' colony conditions also alternate less if they find food in the arm during the previous trial (Gaffen & Davies, 1982; Richman, 1989). The apparatus and procedure of the SAB test used by Wong (1969) is shown in Fig. 7.3.

Wong (1969) wondered why H rats alternated less when hungry than when they were full, and why NH rats showed uniform alternation behaviour under both conditions. He assumed that both groups show the same potential for alternation or novelty-seeking but that the H rats were more adept than NH rats in acquiring a food-directed connection or association that was relevant to their motivational state. Thus H rats did not shift their response toward the novel stimuli on Trial 3 as much as NH rats. More H than NH rats ate the pellets in the Trial 3 goal-box when they were food-deprived, another instance of their more adaptive goal-appropriate behaviour relevant to their motivational state. Williams (1973) interpreted these results in a different manner by suggesting that NH rats may seek novel stimuli more than H rats. The possibility that the more 'emotional' NH rats explore more than H rats under certain circumstances prompted Williams to relate it to Halliday's (1966) hypothesis that exploration may be facilitated by mild fear. This rationale and data leading to this hypothesis is elaborated later in the following section.

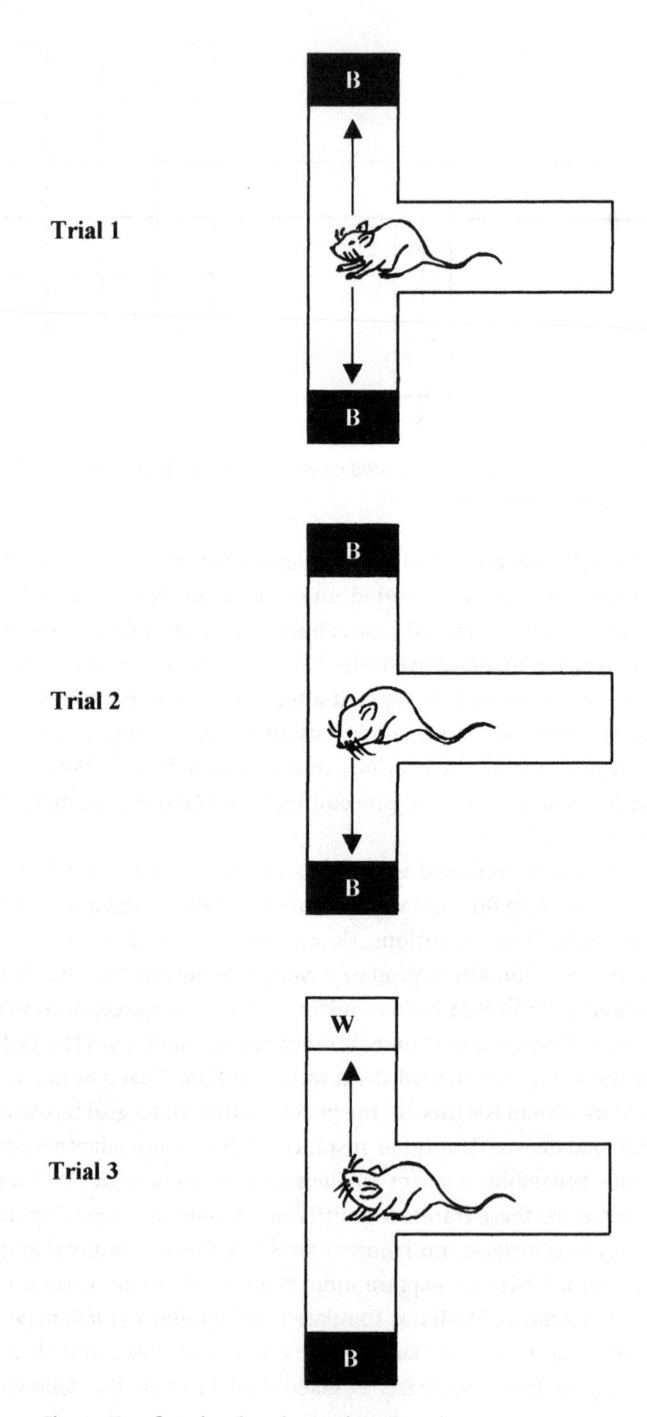

Fig. 7.3 Test for stimulus alternation. (Based on Wong, 1969.)

FEAR AND EXPLORATION

The data indicating that an environmental change can elicit either exploration or withdrawal has led to two sorts of explanation. One based on a *biphasic* system suggests that most novel and complex stimuli activate two motivational systems – one mediating avoidance and the other approach. The observed behaviour is the result of an approach–avoidance conflict, and reflects two separate but interacting motivational systems (Montgomery & Monkman, 1955; Schneirla, 1965; Valle, 1972). Rats of lower emotionality are more likely to react to a novel and complex stimuli with approach than avoidance, whereas those of higher emotionality require a longer period of habituation before such a reaction is manifested. There are handling differences in the level of exploratory behaviour upon the rats' initial exposure to the novel and complex stimuli in the apparatus, but on later trials the differences wash out (Denenberg, 1969).

The other explanation based on *monophasic* theories, in contrast, propose a single motivational process operating with a threshold below which exploration is more likely and above which withdrawal is more likely. The type of arousal theory espoused by Fiske & Maddi (1961) would attribute exploration to small–moderate increases in arousal, and withdrawal to large increases. Whether the underlying motivation is above or below the threshold depends upon various factors, including the magnitude and nature of the environmental change. Hence this theory would predict handling difference when the rats are given a choice between novel/complex stimuli and familiar ones because of the arousal potential differences in these two groups.

Another form of monophasic theory, proposed by Halliday (1966) and Lester (1967) suggests that exploration is actually motivated by the fear generated by environmental changes. They argued that rats tend to explore novel stimuli, not because of an attraction to novelty, but because they arouse low levels of fear. Most stimuli elicit arousal and fear in the rat on their first presentation, and if they are harmless, they will cease to elicit fear and hence exploration. Although the proponents of this theory agree that high levels of fear elicit withdrawal, they also add the assumption that lower levels of fear elicit approach and investigation. This was demonstrated by findings that rats sometimes approach, and have a preference for, distinctive stimuli such as stripes that have previously been associated with mild electric shock (Williams, 1972; Wong & Bowles, 1976). Note here that preference is measured through the rat's *duration of time spent* in the striped arm the day after it was placed in it and given shock or mere placement. They were not tested for preferences of novelty. In contrast the type of experiment in which the animals are placed in an apparatus where they respond for novelty/complexity via the LCBP test or contact of objects in the open field, the experiments influenced by the monophasic theory deals with their preference of stimuli associated with mild shock.

In addition to his earlier result, Williams (1973) found that NH (more emotional) rats not only showed greater initial choice of the striped arm but also spent more time there than the H (less emotional) rats. Wong & Bowles (1976) did not find any differences in initial preference as a result of handling, but NH rats did spend more time in the striped arm than the H rats. This difference was manifested in the form of a statistical interaction between handling and days of testing. The NH rats maintained their level of time spent in the striped arm for three days, while the H rats decreased their exploration of this arm with successive sessions. Following placement and mild shock in the striped arm on Day 4, the NH rats spent *more* time in the striped arm than *non-shocked* NH control rats, but the H rats did *not* show any differential reaction to shock. When tested *without* shock on Day 5, the rats that had been shocked on Day 4 chose and spent more time in the striped arm than those that had not. Handling differences were not evident during this phase of the experiment, and suggests that H rats are just as motivated to seek complexity as the NH rats. The assumption behind both studies is that *mild fear* of certain stimuli motivate exploration of them. However, more intense shocks would result in avoidance of stimuli associated with them (see Chapter 9 for an elaboration of this phenomenon).

Russell (1973) offered an alternative interpretation which suggests that mild shock may function in the same way as any environment that has been associated with an 'interesting' stimulus. The rats may be showing 'anticipatory curiosity'. The area may be relatively incongruous because it was once, but is no longer, associated with such a stimulus. Russell argued that even if shock effects are interpreted in terms of fear facilitating exploration, this need not mean that fear is necessary or sufficient for the occurrence of exploration in general. The positive correlation between indices of fear and measures of exploration need not indicate a causal relationship. Denenberg's (1969) factor analysis of open-field measures revealed that the dimensions of emotionality and stimulus seeking are orthogonal to each other. However, the complex interactions between handling and mild shock effects on stimulus choice and time spent in the arm associated with shock (Wong & Bowles, 1976) indicate the difficulties in making such conceptual distinctions.

Although fear of strong novel stimulation may inhibit exploration of that stimulation, any exploration that does occur will yield information that will reduce fear. Fear *may* facilitate exploration under *special circumstances* as in the Halliday (1966), Williams (1973) and Wong & Bowles (1976) experiments, but these *cannot* be regarded as evidence that attribute *all* exploratory behaviour to this motivational source.

Causal mechanisms of the handling effect

The considerable evidence indicating that early stimulation alters the adult rats' behavioural and physiological reactions to novel stimuli as well as to aversive stimuli that evoke stress led Levine & Mullins (1966) to develop the following hypothesis. They suggest that handling leads to a more rapid intitial response followed by more rapid habituation. The speed and short duration of the response in the H rat serves the purpose of mobilising its resources at the moment when it is under stress. Previous work by Levine's group indicated that infant treatment hastens the maturation of the pituitary–adrenocortical system which is involved in the reaction to stress. This suggests that the physiological chain by which early stimulation affects emotionality and reactions to novel stimuli involves modification in the sensitivity of the adrenocortical system.

When the adult rats were placed in an open field and the amount of corticosterone was assayed following initial exposure and 15 minutes later, the NH rats showed greater corticosterone level than the the H rats (Levine *et al.*, 1967). However, if these rats were shocked, the H rats release a higher level of corticosterone at the time of shock termination but a lower level of this hormone than NH rats over the next 15 minutes (Haltmeyer, Denenberg & Zarrow, 1967). Similar results were found when Pfeifer & Davis (1974) measured the release of a rate-limiting enzyme in the synthesis of adrenaline and noradrenaline among H and NH rats. Taken together, the results of these experiments suggest a more *flexible* and *adaptive* adrenocortical response to stress in the H rat. In the absence of stress, its resting level of the hormone is lower. If a mild stress is applied, it remains low. However, if a more intense stress is applied, it rises rapidly to a level that is higher than in the NH rat, but when this stress is terminated, its adrenocortical activity falls to a relatively low level.

Levine & Mullins (1966) proposed that *variation* in the concentration of adrenocortical steroids during infancy affects the development of the rat infant brain. The release of ACTH from the pituitary, which in turn stimulates the release of glucocorticoid hormones, is under the control of the hypothalmus. The latter is itself influenced by a number of other brain structures such as the hippocampus, which contain glucocorticoid receptors. Glucocorticoid hormones, such as corticosterone released in the adrenal cortex, reach the brain and bind to these receptors. This initiates a chain of events that leads to inhibition of pituitary release of ACTH. Thus a negative feedback loop is formed that restricts further release of corticosterone. This feedback loop causes the return of corticosterone levels to baseline levels upon the termination of a stressor. These events provide an explanation of the effects found in the Haltmeyer *et al.* (1967) experiment.

Support for the Levine & Mullins (1966) hypothesis may be found in experiments by Meaney *et al.* (1985) which demonstrated an increase in the number of glucocorticoid receptors in the adult brain of H rats. The increased density of receptors was found only in the hippocampus and the frontal cortex, regions which are implicated in anxiety. Such results suggest that early stimulation affects adult levels of fear by acting upon the hippocampus and frontal cortex of the developing brain to equip it with more sensitive feedback control over corticosteroid release in response to varied conditions of stress. More recent work from Meaney *et al.* (1993) has detailed specific physiological mechanisms of handling effects and the development of the hypothalamic–pituitary–adrenal responses to stress. They propose that handling increases glucocorticoid receptor gene transcription, thereby increasing sensitivity to glucocorticoid negative feedback regulation. These changes alter the activity within hypothalamic corticotrophin-releasing factor neurons, and these serve to determine neuroendocrine responsivity to stress.

ALTERNATIVE MODELS OF STIMULUS SEEKING

These models provide an explanation that focuses on the cognitive aspects of stimulus change. They pay greater attention to the processes through which animals detect environmental changes, and suggest that responding to such changes must involve some sort of comparator process, with current stimulus input being compared with previous inputs. Salzen (1962, 1970) proposed that a familiar environment is embodied in a set of internal representations or neuronal models that have been built up through experience. A *discrepancy* between a model and current input arising from an environmental change serves to activate a behavioural system that is directed at eliminating discrepancy. Small discrepancies produce low levels of activation and result in approach and investigation of the source of change, which then serve to establish a new model and eliminate the discrepancy. Larger discrepancies, producing more intense activation, lead to withdrawal from the discrepant source. This model considers the role of arousal processes in a different perspective from the Berlyne (1967) or Fiske & Maddi (1961) models discussed in the previous section, but is compatible with them.

The environmental modelling approach is closely related to one that emphasises the concept of a *cognitive map* (Toates, 1983, 1986, 1988). The latter term implies a neural representation of a set of spatial and/or temporal relationships in the environment. Simply by being exposed to stimuli, animals can form internal representations of them. Exploration, such as roaming in an environment or observing stimuli, is behaviour that involves the assimilation of information concerning the form of objects and their spatial lay-out. An early systems approach by Deutsch (1960) contained many of the elements of modern modelling theory. He regarded exploration of an environment as a means of establishing internal representations of significant landmarks or 'links' and associations between them. This relationship defines a cognitive map. The links represent sites of food, water, escape routes and other important sources and cues leading to them. Subsequent efforts by O'Keefe & Nadel (1978) have attempted to identify the hippocampus as the core brain structure involved in environmental mapping. They regard exploration as behaviour that first builds and then revises, cognitive maps. They propose that 'misplace detectors' act to detect disparity between incoming sensory information and the internal representations that the animal possesses for that environment. Although *mismatch* is regarded as a causal factor underlying exploration, they leave room for the exploration of familiar stimuli. When the rat roams about inspecting its territory, if no mismatch occurs, exploration is rapid. If there is a mismatch, it shows a lengthier investigation.

The approach developed by Toates (1983, 1986, 1988) regards the role of exploration as a means of building a model of the world and its causal texture which allows the animal to anticipate remote events. Exploratory behaviour is directed at the acquisition of information as its consequence. Toates favoured an *incentive* model in which exploration is aroused by synergistic properties of external stimuli and internal states. In this respect the analysis of this system is consistent with the general approach arising from discussion of the other motivational systems described in the earlier chapters. Incentive models emphasise the role of environmental stimuli in arousing motivational states (Bindra, 1969, 1974, 1978). Objects in the environment have *potential incentive value* for exploration. Whether they elicit investigation depends upon the internal state of the animal. This state, in turn, is associated with *disparity* between incoming information and internal models of the environment.

Toates (1983) pointed out that 'whereas aspects of exploration can only be understood in terms of a cognitive map, conversely, the construction of a cognitive map is to be understood in terms of the outcome of exploratory behaviour'. When cognitive maps are formed, landmarks are coded along with their relevance to, for example, food, water and aversive stimuli. Exploration serves to first establish and then conform this inner representation. Once this representation is formed, and a biological reinforcer is introduced to the environment, these events are assimilated into the matrix of established representations. Following the construction of the cognitive map, various sites within the map successively form the goal towards which movement is directed. A site is specified as a goal, and it forms the set-point of a negative feedback system (Powers, 1978; Toates, 1983). The animal moves towards the goal as specified by the set-point. Thus food or water deprivation enhances the strength of food-related or water-related goals. Exploration enables such expectations to be constructed and refined. In this respect, exploration appears to facilitate the attainment of other biologically significant incentives. This analysis is relevant to extrinsic exploration, but one must consider intrinsic exploration involving exploratory acts that are not instrumental in achieving any particular goal other than performance of the acts themselves (Berlyne, 1960; Hughes, 1997). O'Keefe & Nadel (1978) claimed that their theory 'accords curiosity, the status of a major motivation, the driving of information incorporation into cognitive maps'. Nevertheless, they also argued that one need not postulate a specific motive for exploration; it is assumed to be aroused by unpredictable external stimuli.

The motivation of intrinsic exploration has been dealt with by proponents of the optimal arousal and fear approach, but its treatment is implicit in the cognitive map modelling theories which focus on the possible ways in which inner representation of objects and events are manifested. However, Hughes (1997) suggested that as no single approach has an adequate explanatory or predictive power, we should just accept the possibility that animals have some type of 'need' for sensory change, which can be satisfied by intrinsic exploration. He also felt that there are difficulties in distinguishing between extrinsic and intrinsic exploration with activity measures, and that tests of free exploration involving measures of active choice of stimuli are preferable. I have alluded to similar methodological issues in the earlier discussion on disentangling the effects of handling on activity level and stimulus seeking tendencies.

The *cognitive incentive* view of exploration does *not* deal with the effects of individual differences in emotional reactivity as they affect exploratory behaviour directly, as does the *optimal arousal* view. Although environmental stimuli arouse motivational states, there are differences in the arousability of individual animals, many of which are due to variables such as early rearing, types and level of stimulation during infancy, etc. Yet one must acknowledge that the *optimal arousal* position has difficulties in explaining the rat's choice or preference of arms leading to stimuli associated with mild electric shock (Halliday, 1966; Lester, 1967; Williams, 1972; Wong & Bowles, 1976). Toates (1983) proposed that the rat's exploration is prompted by a significant event, namely shock. Russell (1983) explained such results by assuming that the shocked rats have a greater need for information about these stimuli and thus explore them more. He posits that a stimulus previously but no longer associated with stressful stimulation will have greater incongruity than it will for an animal which experienced the same stimulus without such a history. The rat explores despite a small amount of fear because the stimulus is more incongruous.

PHYSIOLOGICAL MECHANISMS OF SPONTANEOUS ALTERNATION BEHAVIOUR

It follows from our discussion and analysis of tasks used to measure stimulus seeking behaviour that tests involving ambulation scores as well as response rate do not provide clear indicants of this disposition as do choice behaviour measures. An examination of the literature on physiological determinants of exploratory behaviour indicates that most of the experiments have focused on lesion effects on alternation behaviour, partly because of the simplicity of this method. In general, alternation is affected by alterations in the neural tissue in the limbic system, particularly in the hippocampal area. Prior to hippocampal lesions, rats show above-chance levels of maze alternation, but after lesions, alternation is decreased by massive bilateral lesions (Dalland, 1970; Kirkby *et al.*, 1967; Stevens & Cowey, 1973).

Douglas's (1989) review led to the conclusion that the hippocampal system is crucial to SAB, and that this structure mediates Pavlovian internal inhibition. His rule of thumb for understanding this analysis is that if inhibition facilitates performance of a task, then rats with hippocampal lesions will do poorly. If the task does not involve inhibition, the lesioned animals will perform normally. According to Pavlovian inhibition theory, SAB reflects the effects of habituation. In intact animals, alternation occurs as a result of internal inhibition arising from stimulus satiation or habituation of the familiar stimulus. The hippocampal lesion produces a weaker build-up of inhibition, and this results in a decrease in alternation in the animal. The factors responsible for the hippocampectomised rat's first choice would also operate on the second trial and are relatively uninfluenced by the first choice. This would tend to cause the second choice to be the same as the first.

Douglas (1989) pointed out SAB is one of a handful of behaviours that are hippocampus-dependent. Other tasks affected by hippocampal lesions include the radial maze, and differential reinforcement of low rate delayed lever-pressing. As well as modulating SAB the hippocampus is implicated in stimulus-seeking activity in general. Gray (1987) examined the function of the hippocampus from a different perspective and proposed that the hippocampus is also involved in the analysis of novelty by the organism. During periods when the organism is aroused, the cortex and a lower brain region – the reticular formation – respond with high frequency, desynchronised electrical rhythms. When this occurs, electrical recordings from the hippocampus indicate the emission of a regular 4–9 Hz pattern known as a *theta wave*. The hippocampal *theta rhythm* is a prominent response to novel stimuli as well as to cues associated with painful stimuli and frustration (details are described in Chapter 8). Recent work suggests that these states are related in that *novelty stress* induces theta waves in rabbits (Yamamoto, 1998). The concept of novelty stress was proposed by Katz (1988) and van Aberleen (1989) to explain the release of adrenal stress hormones found in rats following exposure to novel stimuli. Earlier work by Routtenberg (1968) indicated that a theta rhythm was induced when rats were exposed to a novel environment but the theta wave decreased during repeated presentations of that environment. He hypothesised that hippocampal theta activity is related to attention mechanisms and information processing. A behavioural study by Raphelson, Isaacson & Douglas (1965) indicated the role of the hippocampus in reactions to novel stimuli. The experimenters rewarded rats in the runway with food and introduced a novel stimulus while they were running. The control rats slowed down when presented with this stimulus, but hippocampectomised rats fail to show this decline. Such results can be interpreted in terms of the *inhibitory control system* of the hippocampus.

Ontogeny of mechanisms of spontaneous alternation behaviour

An extensive review by Spear & Miller (1989) indicated overwhelming evidence of SAB changing dramatically with postnatal development.

This fact raises the question of what it is about infants that cause them to alternate less than adults. One possibility is that infants are more fearful of new stimuli than adults. Spear & Miller (1989) regard it almost as a truism that when the emotional state of the rat fits the characteristics consistent with the term 'fear', less SAB is exhibited. However, the results from handling and SAB by Wong (1969) as well as those on handling and stimulus preference by Williams (1973) indicate specific conditions where this generalisation might be qualified. Other studies concerned with age differences in SAB demonstrate that this behaviour is low during the fourth postnatal week and rises thereafter (Bronstein *et al.*, 1974; Hess & Blozovski, 1987). The latter suggest that there is a progressive increase in fear manifested by defecation and vocalisations in the experimental apparatus between postnatal days 20 and 30, and a subsequent decrease that corresponds, respectively with the decrease and subsequent increase in alternation during this period.

Infant rats may require a longer period of stimulus exposure than adults to habituate to that stimulus. If the event to be habituated is an object or location that can be approached by the animal from different perspectives, then the habituation rate and SAB are related to age. Feigley *et al.* (1972) and Parsons, Fagan & Spear (1973) found that 15-day-old rats require a longer period of exposure than older rats to habituate to a nose-poking response to a hole in the wall. In other words, it takes longer for the 'new' objects to become 'old' among younger than for older rats. Infant rats also are more likely to forget where they have been from one trial to the next (Spear, 1978). Thus the ephemeral memory for events among infant rats will lead to less SAB. At the physiological level, the hippocampus seems to mature at a rate that is correlated with age-related changes in SAB. Brain structures important for SAB are functionally immature in infant rats. The emergence of SAB parallels the development of the hippocampus (Altman & Das, 1965).

NEOPHILIA AND NEOPHOBIA

In their review of ethological studies on rodents, Barnett & Cowan (1976) described the animal's tendency to explore a new place when given access to it as *neophilia*. The material examined in the early part of this chapter focused on theories and experiments dealing with the mechanisms of it. The function of neophilia is assumed to promote dispersion. The animal's chances of finding necessities are believed to be enhanced. The authors also pointed out that commensal species (i.e. those that are dependent upon humans for food) manifest a tendency to avoid new objects, or *neophobia*. Because pests have been and are still being subjected to the selective pressures of human predation, Barnett & Cowan (1976) hypothesised that these animals have undergone behavioural changes that may protect them from the consequences of neophilia. Neophobic reactions are believed to be responsible for the difficulties in trapping rats or getting them to eat poison bait (Chitty, 1954). Researchers have postulated dual neophilic–neophobic processes to explain whether particular animals explore or avoid new places or new objects. This perspective has been taken by Barnett (1956, 1958) and Mitchell (1976) in their studies, indicating greater neophobia among wild rats than among laboratory rats. Mitchell attributed this effect to emotional differences between wild and domestic rats. More recently, Timmermans, Roder & Hunting (1986), Timmermans *et al.* (1994) and Vochteloo *et al.* (1991) found that the acquisition of phobias to harmless objects by New World cynomolgus monkeys depended partly on rearing conditions (i.e. surrogate or own mother) as well as innate disposition.

Neophobic tendencies are present in the domestic rat population as well as in those in the wild. Although there is greater variability of this trait among the former, individual differences in response to novelty is transituational (Corey, 1978). A recent experiment by Minor *et al.* (1994) found that individual differences in open field exploration predicted neophobic responses to a novel saccharin solution in the rat's home cage. Food neophobia, the reluctance of animals to consume novel food, has been extensively studied (see review in Wong, 1995). Ominivorous species such as humans and rats, face a dilemma when they encounter new foods. There is benefit to express a neophilic tendency to approach new food because it is a potential source of nutrition, but there is also a potential cost because the food may be toxic. This neophilic–neophobic conflict is referred to as the *omnivore's dilemma* (Rozin, 1977). Aspects of this dilemma were discussed in our examination of food selection mechanisms in Chapter 5. The idea that food neophobia is a more general global process that regulates responses to novelty was suggested by Minor *et al.* (1994) in their interpretation of their data.

Although earlier studies led researchers to regard neophobia as a reaction shown only in wild rats, subsequent experimenters have attempted to demonstrate this phenomenon in domestic rats under specific test conditions. Wong's (1995) review of the literature indicates that most of these experiments involved tests in which the rat's consumption of novel-tasting fluids was compared with that of an unflavoured and the familiar fluid (water). In general, these experiments indicated manifestation of flavour neophobia, although the phenomenon was enhanced when illness-inducing toxic injections were administered to the animals prior to their exposure to the novel-tasting edible. A comparative study by Miller & Holzman (1981) revealed differential neophobic reactions of rats, guinea pigs and gerbils when tested with solutions containing highly dissimilar flavours. This led Wong & McBride (1993) to use solid foods such as nuts or seeds as the test stimuli, as well as to measure response latency rather than consumption in their studies on non-deprived gerbils and hamsters. Neophobia is a short-lived phenomenon and is best detected during a brief test (Carroll *et al.*, 1975). Both species showed a longer latency (neophobic reaction) to salty, sweet and sour flavours compared to familiar unflavoured seeds or nuts.

Food neophobia has also been studied in humans (see review by Pliner & Pelchat, 1991). Children begin to reject a variety of foods at about two years of age (Birch, 1990) – an age at which familiar foods are strongly preferred. Studies have shown that exposure to foods increases preferences for them (Birch & Marlin, 1982; Birch *et al.*, 1987; Pliner, 1982). This is consistent with the results of Corey's (1978) summary of animal studies showing that exposure to variety increases approach to novelty under diverse circumstances later in life. For example, chimpanzees (Menzel, Davenport & Rogers, 1961) and rats (Sheldon, 1969) show facilitation of approach to novel objects as a result of experience with novelty. Similarly, pre-exposing rats to several different flavours reduced neophobic responses to a novel test flavour (Braveman & Jarvis, 1978; Capretta, Petersik & Stewart, 1975). Modulation of neophobia in humans is also influenced by experience. The number of foods tried in the past is predictive of willingness to try new foods (Frank & Raudenbush, 1996), and neophobia decreases with age (Koivisto & Sjoden, 1996; Pelchat & Pliner, 1995; Pliner & Hobden, 1992). This interpretation is based on the assumption that experience with novelty enhances approach to novelty.

One wonders whether diverse experiences in general would influence food neophobia. Would experience with novelty in one domain transfer to approach in other domains? Franchina & Dyer (1989) studied cross-modality transfer in chicks and concluded from their results that novelty is a general stimulus property that can be separated from the specific characteristics of the stimulus. Results of experiments by Hennessy *et al.* (1976) and Hennessy, Smotherman & Levine (1977) also support this conclusion. The hypothesis of cross-modality novelty seeking in humans is supported by Raudenbush & Frank's (1999) finding of significant correlations between the number of foods and activities tried, and the willingness to try food and activities.

Although food neophobia has been clearly established in numerous studies, the understanding of the physiological mechanisms underlying it is less substantial. Corey (1978) conceptualised neophobia as a specific instance of *neotic behaviour* – a term referring to a range of responses to novel stimuli encountered in the animal's external as well as internal environment. In his studies on factors influencing open-field behaviour in rats, Corey implicated stress arising from a heightened arousal in this apparatus as a factor that elicits neophobic behaviour. A subsequent experiment by Lee, Teng & Chai (1987) suggested that the open-field activity of mice was potentiated by stress, but that such activity was attenuated by antagonists of endogenous stress hormones.

In his research on the effects of endorphins on exploratory activity, Katz (1988) suggested that rodents encountering novel stimuli experience a stress reaction and that the hormones released during this stress response mediate the neotic behaviour. Van Aberleen (1989) also proposed that the behaviours occurring when rodents encounter new situations may involve a certain degree of fearfulness. The physiological mechanisms mediating reactions to novel stimuli in an unfamiliar environmental setting may also be involved when the animal is in contact with novel foods. Thus an animal's initial hesitation to explore novel stimuli in its setting may be mediated by a mechanism which also causes it to be hesitant during its first contact and subsequent ingestion of a novel food. According to this hypothesis, flavour neophobia would be attenuated by an anti-anxiety agent if stress variables are involved in the manifestation of this phenomenon. Anxiolytic (anti-anxiety) agents alter the peptide hormones that mediate stress and also inhibit neotic behaviour in the open field. Shepherd & Estall (1984a, b) and Fletcher & Davies (1990) found that the anxiolytic drug, chlorodiazepoxide, decreased the rats' eating latency of novel food relative to that of non-treated animals. This drug also increased the amount of time that the rats ate the novel food but not the amount eaten. Lao and Wong (unpublished data) studied the effects of another anxiolytic drug, diazepam, on the hamster's response latency to a novel flavoured edible. The results with this measure of neophobia indicate that those injected with diazepam prior to the test with a novel sweetened sunflower seed showed faster response latencies than control animals who received isotonic saline injections.

Another approach to test the novelty stress hypothesis is to study the effects of infant handling on neophobia. As discussed earlier it was noted that studies indicating that H rats show less elevation of plasma corticosterone in response to novel stimuli than NH rats (Levine *et al.*, 1967). With these considerations in mind, Weinberg, Smotherland & Levine (1978) predicted that flavour neophobic responses would be attenuated by infant handling. When presented with a pleasant (sweet) novel solution, H rats drank more of it than NH rats. These results are consistent with the notion that exposure increases subsequent intake because it reduces the novelty of the substance and thereby attenuates the neophobic or aversive reactions elicited by the novelty. It is also possible to test this hypothesis by manipulating brain structures such as the hippocampus and amygdala that have been implicated in novelty-induced responses (Nachmann, Rauschenberger & Ashe, 1977; van Abeleen, 1989). The hippocampus is implicated as a crucial structure for stimulus seeking behaviour in general (Douglas, 1989; Gray, 1971), and recently, Yamamoto (1998) demonstrated that novelty stress induces hippocampal theta waves in rabbits. Maren *et al.* (1994) found that emergence neophobia in rats correlates with hippocampal receptor binding. The amygdala is implicated in neophobia in that rats with bilateral lesions of the basolateral region attenuate their usual tendency to restrain consumption of the novel liquid (Borsini & Rolls, 1984; Kesner, Berman & Tardif, 1992; Kolakowska, Larue-Achagiotis & Le Magnen, 1984).

SENSATION SEEKING IN HUMANS

Some of the ideas derived from animal studies described in this chapter are relevant to the analysis of sensation seeking in humans. Zuckerman (1994) defined this trait as characterised by the seeking of varied, novel, complex and intense sensations and experiences. His theory began with the hypothesis of consistent individual differences in optimal levels of stimulation and arousal, expressed in certain types of human activities and measured with a self-report questionnaire. Zuckerman (1984) proposed that genetic programmes influencing the biochemistry of the nervous system provide the biological basis for the trait. Differences in activity of the brain catecholamine systems influence arousability of the cortical areas. Although environment may determine the particular forms of expression of the trait, the amount of variation in stimulation received during infancy and early childhood also influences the developing trait. Sensation seeking is seen as the outcome of a conflict between states of anxiety and sensation that vary as a function of novelty and appraised risk. Note the parallel between this explanation of human sensation seeking with the biphasic model of fear and exploration discussed earlier in this chapter. Zuckerman (1994) suggested that the brain of the high sensation seeker may be more accessible to novel or intense stimuli. Our earlier examination of early stimulation effects also leads to a similar notion with respect to H rats.

Zuckerman (1994) regarded the sensation seeking trait as one of adaptive significance. In obtaining food, or mates, or competing with others for these goals, the animal or human must leave the shelter of its domicile and approach objects in a potentially dangerous environment.

This trait represents an optimistic tendency to approach novel stimuli and explore the environment. Persons low on the trait tend to withdraw under circumstances where the stimuli are too novel or the outcome too uncertain. Dopamine in the nigrostriatal system energises active exploration of the social and physical environment, and it provides the positive arousal and reward associated with novel and intense stimulation. Serotonin in the pathway originating in the medial raphe system is involved in the inhibition of behaviour in the presence of novel or threatening stimuli. In this respect, Zuckerman regarded dopamine as the drive mechanism and serotonin as the brakes of the approach mechanism for sensation seeking behaviour. Support for Zuckerman's (1994) theory was documented by his detailed examination and discussion of data on risk-taking, driving habits, health, gambling, financial activities, alcohol and drug use, sexual behaviour and sports.

SUMMARY

Exploratory activities enable animals to acquire information about the environment which is processed in the nervous system during contact with it. Both stimulus variation and stimulus intensity contribute to the *impact* of a stimulus, and the amount of impact sought depends upon the animal's characteristic level of arousal. Early rearing conditions influence an animal's basal arousal level and also its stimulus seeking tendencies. Environmental change or variation can elicit either exploration or withdrawal. The *two-factor* explanation suggests that the observed behaviour to novel and complex stimuli reflects the effects of two separate but interacting systems, one mediating *avoidance* and the other, *approach*. The monophasic explanation proposes a single motivational process operating with a threshold, below which exploration is more likely and above which withdrawal is more likely.

Cognitive explanations of stimulus seeking focus on the animal's internal representation of its environment. Its response to environmental change involves a comparator process in which current input is compared with previous inputs. A discrepancy between a model and current input activates a behavioural system that is directed at eliminating discrepancy. The cognitive approach regards exploration as a means of building a model of the world and and its causal texture. Objects in the environment have a 'potential incentive value' for exploration. Whether they elicit investigation depends upon the internal state of the animals, which, in turn, is associated with the disparity between incoming information and internal models of the environment.

8

Aversive motivation systems: fear, frustration and aggression

When animals are exposed to aversive stimuli, particularly those of pain, they are likely to respond by either withdrawing from or attacking the source of the stimuli. 'Pain is an anatomically developed sensory system genetically differentiated for survival and the defence of the body. Responses to painful stimuli either involve the skeletal musculature or are internal but they are experimentally quantified as escape and avoidance' (Le Magnen, 1998, p. 4). Both types of responses to this source serve adaptive functions. In many species, specific escape mechanisms have evolved for dealing with physical danger. One of the simplest is the withdrawal reflex that removes the organism from damaging stimuli. When specific taste receptors are in contact with bitter substances, they result in a spitting reflex that protects the organism from ingesting possibly toxic substances that are generally associated with the bitter taste. In Chapter 5 the mechanisms by which rats learn to avoid smells and tastes that have previously been followed by illness were examined. Animals will also react to aversive stimuli with attack behaviour, particularly when escape is difficult. Such a reaction is particularly evident in feral animals.

TWO-FACTOR THEORY OF AVOIDANCE BEHAVIOUR

The standard situation or apparatus used to study responses to aversive stimuli is one in which a rat is placed in a shuttle box – a long narrow box divided in half by a partition. The floor of the box is a grid of steel rods that can can deliver a painful shock when activated by electricity. The rules of the experiment are as follows. The rat has a few seconds to cross the barrier over to the other side of the box. If the rat responds in this fashion, the buzzer goes off and nothing else happens. If the rat fails to cross in time, then an electric shock is experienced and continues until the crossing occurs.

In such an experiment the rat is likely not to respond during the first few trials when the buzzer comes on. The electric shock comes a few seconds later, and causes the rat to run and eventually cross over the barrier. Crossing the barrier turns the electrical current off, and this response indicates that the rat has escaped the painful shock. Over successive trials the escape response becomes progressively more likely. Although the rat may freeze or crouch when the buzzer is sounded, after the electric shock begins, the rat quickly makes the crossing-over response that turns off the electric shock. Eventually most rats will display the avoidance responses, that is, they run during the warning interval between the onset of the buzzer and that of the electric shock. This act prevents the shock from occurring, and eventually the animal will continue trial after trial, crossing over to the other side upon the onset of the signal, and thus never experiencing an electric shock.

The explanation of the rat's escape response in which it runs and thus terminates the electric shock following its response poses no particular problem. The electric shock is painful, running turns it off, and so running is rewarded and strengthened. Running is reinforced by escape from pain. We face a more difficult problem in explaining the process that reinforces the avoidance response, thus preventing the electric shock from occurring. If the absence of shock maintains continuous response, this poses the puzzle of how the absence of an event can be a reinforcer. How can a non-event strengthen and maintain behaviour? One solution is to demonstrate that avoidance behaviour is really a response to a stimulus and not to a non-stimulus. This type of explanation forms the basis of the two-factor theory of avoidance behaviour (Miller, 1948; Mowrer, 1960) which accounts for its acquisition in the following manner. First, the rat acquires a conditioned response, namely fear, to the buzzer or other warning signals. Second, the avoidance response reduces the fear. Thus the reinforcement for avoidance responding is not a non-event such as the absence of an electric shock, but fear reduction, an event.

This theory suggests a closer examination of events occurring during the early trials. The buzzer comes on and a few seconds later, the painful shock follows. The latter will produce many stress reactions such as tension in the muscles, increase in heart rate as well as breathing rate, and secretions of hormones from the adrenal gland. These reactions occur before the onset of the buzzer, but eventually, these stress responses begin to occur in response to the buzzer. This constitutes the situation under which classical or Pavlovian conditioning occurs. This is the first of the two factors in the two-factor theory. Stress responses such as muscle tension, rise in blood pressure and heart rate are, in turn, assumed to comprise effects that are unpleasant. These reactions may be regarded as the basis of an emotional state, that of fear. From this perspective, fear is not a mental state but an internal stimulus situation comprising a set of internal stimuli produced by the conditioned stress responses. It is also as unpleasant or aversive as the electric shock. Therefore, any response that terminates fear will be reinforced and strengthened. If the rat crosses over before the shock is administered and it also turns off the buzzer, the stimulus for fear is removed, and so the fear state ends. These events account for the reinforcement of the avoidance response. Fear reduction constitutes the second of the two factors.

In essence, a negative-feedback loop is involved in which a stimulus (the buzzer) produces fear, which in turn, calls forth an action such as running that removes the fear by removing the stimulus that causes it. In this analysis, the avoidance response is treated not as an anticipatory response to something that has yet to happen, but as a response to something that has happened, namely fear produced by the buzzer. This analysis also treats the avoidance response as an event that is reinforced by something that happens, the reduction of fear, rather than as the result of something that does not happen. This type of explanation has been extended to explain irrational forms of human behaviour such as compulsions. Such behaviour is assumed to be motivated by fear, and maintained by fear reduction. The fear itself may be traced to certain experiences that created them, much in the way in which the electric shock creates fear of the buzzer in the rat.

Problems with two-factor theory

Research on dogs has indicated that early in avoidance training, the buzzer provokes fear, but later in training when the avoidance response has been well developed, the animal shows little evidence of fear (Solomon & Wynne, 1954), yet continues to make this response when the buzzer sounds. Subsequent studies indicated little evidence of a causal relationship between physiological measures of fear such as heart rate and changes in defensive responses (de Toledo & Black, 1966; Powell *et al.*, 1971). There is another problem with the two-factor theory. In setting up an avoidance conditioning experiment, the response chosen by the experimenter as the correct one is arbitrary. According to the two-factor theory, any response may be chosen and the animal can learn to escape or avoid the aversive stimuli associated with these acts. However, some responses are learned much more easily than others. These constraints intrigued Bolles (1970), and led to a reformulation of the theory explaining avoidance behaviour. He argued that although a pigeon may readily learn to jump across a treadle to avoid an electric shock, it learns to peck an illumination key in the apparatus to avoid getting an electric shock with great difficulty, if at all. The key-pecking response itself is not difficult for the pigeon if it is reinforced with food for doing so. But it has great difficulty in ever learning this response if it is reinforced through shock avoidance. A similar observation has been made on rats that learn to run in avoidance situations much more easily than they learn to press the lever.

The biological explanation of the constraints on avoidance behaviour proposed by Bolles (1970) points out the following. An animal comes to the experiment equipped with specific ways of responding to danger. These form a hierarchical pattern of innate reactions when the defensive motivation system is activated. Various response patterns, which are components of the system, consist of running away, freezing or attacking. The specific actions included in the system depends upon the species. Tortoises would withdraw into their shells rather than run, and birds fly away rather than run away. Thus, Bolles described and conceptualised these action patterns as *species-specific defensive reactions* or SSDRs, behavioural adaptations that have evolved through natural selection as defence mechanisms.

In explaining the behaviours that animals manifest in the avoidance task, Bolles (1970) argued that animals use instinctive responses in dealing with the situation. If one of them is effective, then that response is learned very quickly. If none of the SSDRs in the animal's repertoire work, then, and only then, will the animals try a response that is not a SSDR. Under these circumstances such a response will be learned slowly, laboriously, and perhaps not at all. When faced with danger, a rat is likely to freeze. If freezing does not work, a SSDR such as running away may be tried, and such a response will be effective in the shuttle box. However, if the rat is tested in a box where a lever-pressing response is required, none of its SSDRs will provide a successful solution to the problem. Although the lever-press response with its paws is easily acquired when the outcome is food because of the association between scratching and the discovery of grain or seeds, it is not part of the rat's instinctive repertoire of reactions to danger.

Bolles's (1970) conception of avoidance is very different from the one offered by the two-factor theory. From the former's perspective, the successful escape or avoidance of shock does not strengthen a new response, such as running from one place to another. The rat already is programmed to know how to run when threatened with danger. What may be learned is that this situation is dangerous, and that a specific member of the class of SSDRs is effective. Defensive behaviour is situation-specific, being confined primarily to dangerous settings, and it is diverse in its appearance such as freezing, flight and fight.

PAVLOVIAN CONDITIONING, ACQUIRED MOTIVATION, AND THE NERVOUS SYSTEM

Although Bolles's (1970) analysis of avoidance behaviour is a viable alternative to the two-factor conditioned fear explanation, analyses which take physiological factors into account are worth examining. One in particular is a model by Gray (1971, 1982, 1987) which incorporates material from neurophysiology and psychopharmacology. Gray pointed out that animals respond differently to conditioned and unconditioned aversive stimuli. When confronted with painful electric shock, animals will typically show increased activity, run, jump, scream, hiss or attack a suitable target such as another animal. However, they respond to a stimulus associated with shock by freezing and remaining silent. Gray suggested that the brain mechanisms mediating these two kinds of reactions are distinct, as are the drugs to which they are sensitive.

Gray (1971, 1987) categorised the mechanism that reacts to unconditioned aversive stimuli as belonging to the 'fight/flight system', and the mechanism that reacts to conditioned aversive stimuli as the 'behavioural inhibition system'. In his analysis, Gray (1987) incorporated the notion, from the two-factor theory, that the aversive nature of stimuli, which have been followed by pain, is a necessary step in the acquisition of an active avoidance response. In addition, Gray drew upon research on 'electrical stimulation of the brain' (Olds & Olds, 1965) indicating that the existence in the brain of two fundamental motivational systems – a 'reward' mechanism and a 'punishment' mechanism. The former mechanism is constructed via connections with the animal's motor system to maximise stimuli that regularly precede the occurrence of its stimulation. In a similar manner, stimuli that regularly precede the occurrence of a punishment acquire, through Pavlovian conditioning, the ability to activate the punishment mechanism. Gray postulated that within the brain there is a 'comparator for punishment' which processes the consequences of its response, and through this mechanism the animal learns to approach safety signals associated with the omission of anticipated punishment. This comparator provides information that indicates the mismatch between expected and received punishment.

FEAR AND FRUSTRATION AS AVERSIVE STATES

Gray's (1987) model indicates the parallel between the effects of punishment and those of the omission of anticipated reward. He argued that 'frustration', a state which is caused by the omission of reward, is functionally and physiologically very similar to the state of 'fear'. However, these states vary in intensity, and the intensity of the reaction aroused by 'frustrative non-reward' is usually less than that aroused by punishment. Gray explains the manner in which frustrative non-reward shares common aversive properties with punishment. An experimenter subjects an animal to frustrative non-reward by first rewarding it for performing some response on a number of trials, and then testing it on trials where reward is omitted. During succeeding test disinclination to perform this response because of the aversive properties of stimuli associated with frustrative non-reward.

Most experiments have shown that the rate of extinction or response decrement resulting from non-reward is related to the magnitude of the reward received by the animal prior to the test trials. In general, resistance to extinction is decreased by increments in the magnitude of reward. That is, large rewards lead to less resistance to reward than a small reward. This is because animals trained with a large magnitude of reward exhibit a greater degree of aversive reactions to non-reward than do those trained with smaller magnitudes. Such an explanation suggests that responding in extinction does not merely decline passively, but rather is actively suppressed by the aversive property of non-rewarded responding.

Wagner (1963) provided a more direct demonstration of the aversive properties of frustrative non-reward. He trained rats to run down an alley, but rewarded them on 50 per cent of their trials – a procedure referred to as a 'partial reinforcement schedule'. On the trials when they were not rewarded, a distinctive stimulus (sound and light) was presented to them as they entered the goal box. They were then tested in a shuttlebox, and when they jumped across the barrier, the stimulus was turned off. The results indicated that they jumped across the barrier more often than the control rats which had experienced the same stimulus in the runway, but not in association with the non-rewarded trials. Such results suggest that stimuli associated with frustrative non-reward are aversive in a similar manner as stimuli associated with punishment.

Consider studies on the effects of a partial reinforcement (PRF) schedule compared with a continuous reinforcement (CRF) schedule involving the same number of trials on animals that were subsequently tested under conditions where reward is omitted altogether. A universal finding is that animals that are trained under the PRF schedule show greater resistance to extinction than those under the CRF schedule. Amsel (1962) developed the classic explanation of the partial reinforcement extinction effect (PREE) in terms of frustrative non-reward mechanisms. His theory suggests that, as a result of non-rewarded trials, an animal on a PRF schedule comes to experience conditioned frustration during its performance of the instrumental response. This state of frustration sets up distinctive internal stimuli, which come to act as cues to continue performing the instrumental response, given that these cues are followed by a reward on the rewarded trials. Thus, the stimulus feedback from the internal state of frustration becomes a signal to the animal to maintain persistence of the formerly reinforced response. When reward is terminated altogether, the resulting frustration is not new but has been associated with non-rewarded responding in the past, and thus responding persists. By contrast, rats that have always experienced reward during training now encounter their first experience of non-reward during the extinction test trials. To the extent that non-reward is aversive, responding is eliminated swiftly because of competing responses elicited by frustration.

Just as a PRF schedule is a method of training an animal to continue responding despite non-rewarded encounters, it is possible to keep an animal responding in spite of punishment if a similar experimental arrangement is made. That is, by introducing an electric shock initially at a very low intensity and then gradually increasing its intensity and always following it with reward, the animal's resistance to punishment can be thereby increased. Such an analysis was developed by Brown & Wagner (1964). It led them to conduct an experiment designed to show the similar behavioural effects produced by non-reward and punishment. They trained three groups of rats: one on a CRF schedule (Group C); one on a PRF schedule in which 50 per cent of the running responses were rewarded with food (Group PR); and a third group on a special punishment schedule (Group P). The third group received exposure to food and electric shock after they entered the goal box. Food was available during every trial, and the shocks were administered on 50 per cent of the acquisition trials.

Following training, each of the three groups was divided into two subgroups for testing, one under consistent non-reward and the other under a schedule wherein food and electric shock were present on each trial. The results indicated that exposure to either non-reward or punishment during acquisition produced resistance to any of the decremental effects of the test events. Group P rats tested under punishment and Group PR rats tested under non-reward showed little decrease in responding. As predicted by the fear = frustration hypothesis, the Group PR subjects were less disrupted by punishment and Group P rats were less disrupted by non-reward than were the corresponding Group C subjects. These results indicate that partial reinforcement and gradually increasing punishment may train subjects to continue responding in the presence of aversive cues that are different from those encountered initially. Thus tolerance for non-reward carries with it tolerance for punishment and vice versa.

Drug studies on reactions of aversive stimuli

The results of the Brown & Wagner (1964) study indicate functional similarity between aversive stimuli arising from punishment and frustrative events. There are also physiological similarities underlying these functional effects. Alcohol can affect an animal's behaviour in an approach–avoidance conflict situation in which a food-deprived rat is trained to run to food in a goal box. When it enters the box it receives an electric shock and is thus punished. If the shock intensity is set at a level that just prevents the rat from entering the box, an injection of alcohol will enable the rat to maintain running behaviour (Miller, 1959).

If there is a physiological overlap between the systems activated by punishment and by frustration, alcohol should also reduce frustrative reactions. This hypothesis can be tested by studying rats injected with alcohol during their acquisition of a response under the PRF schedule. Frustration theory suggests that rats on a PRF schedule learn to use internal cues of non-reward as cues for continued approach behaviour. If frustration is reduced, by administering the rats with drugs during training, certain results should follow when they are tested without the drug in the extinction trials. The usual difference between PRF and CRF groups should be abolished. This prediction was confirmed in experiments using sodium amytal by Ison & Pennes (1969) and Gray (1969). Identical results were found by Feldon & Gray (1981) using a different fear-reducing drug, chlordiazepoxide (Librium). Because rats administered with these drugs during training were prevented from developing tolerance of frustration, they were more impeded by the aversive effects of frustrative non-reward during the extinction trials than were control rats that acquired the response under non-drug conditions.

Neural structures and the behavioural inhibition system

The limbic system, a series of interlinked nuclei and tracts which surround the thalamus and lie beneath the cortex, is implicated in the regulation of emotional reactions (MacLean, 1970). The hippocampus is a structure within this system and has been implicated in the analysis of novelty by the organism (Gray, 1971) as well as in the inhibition of responses to cues associated with aversive stimuli (Douglas, 1967; Kimble, 1968). During periods of arousal when the cortex and the reticular arousal system register desynchronised electrical rhythms, cells in the hippocampal region display a narrow range frequency of firing from 4 to 9 Hz. This pattern is referred to as the *hippocampal theta wave*. One of the functions of the hippocampus is to inhibit the activating systems of the lower brain upon repeated presentations of the stimulus (Kimble, 1968). This function enables the animal to decouple its attention to new or more consequential environmental events. The hippocampal theta rhythm is blocked if an analgesic drug is injected into the animal, and its responsiveness to painful stimuli is simultaneously decreased (Gray, 1970). Such results prompted Gray to propose that there is a single physiological system that mediates the effects of punishment and frustrative non-reward, in addition to its relationship to processing of novel stimuli. Gray (1976) regards the physiological mechanisms mediating these effects as the structural basis of a behavioural inhibition system (BIS) which is antagonised by analgesics and barbiturate drugs.

The septal area of the limbic system has two-way connections with the hippocampus and appears to lie astride the input for pain and novelty to the latter structure. Lesions in the medial nuclei of the septum abolish the hippocampal theta wave when the animal is presented with fear-related and novel stimuli. Fig. 8.1 illustrates the effects of such lesions on hippocampal function. It has also been observed that neurons in the nuclei fire in bursts in phase with the theta rhythm. Gray (1971) was able to drive the hippocampal theta rhythm through electrodes implanted in the septal area at frequencies lying within the natural theta range. This led Gray to conclude that the septal area contains the pacemaker cells for the hippocampal theta wave. Following further research on this system, Gray (1987) suggested that it has the general task of comparing actual with expected stimuli. If actual stimuli are successfully matched with expected ones, it functions in a 'checking mode' and behavioural control rests with the other brain

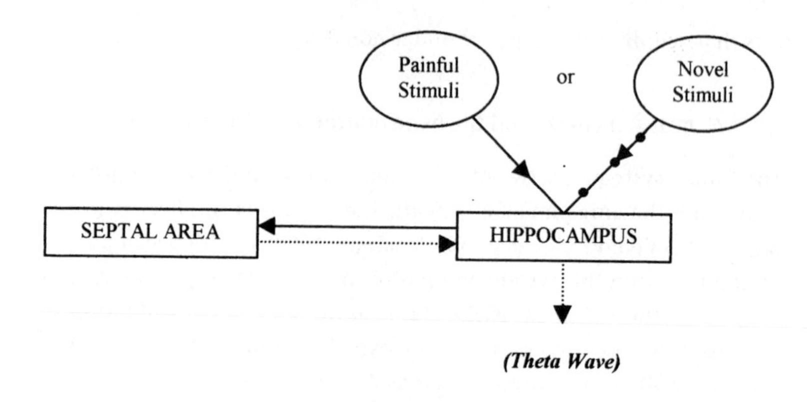

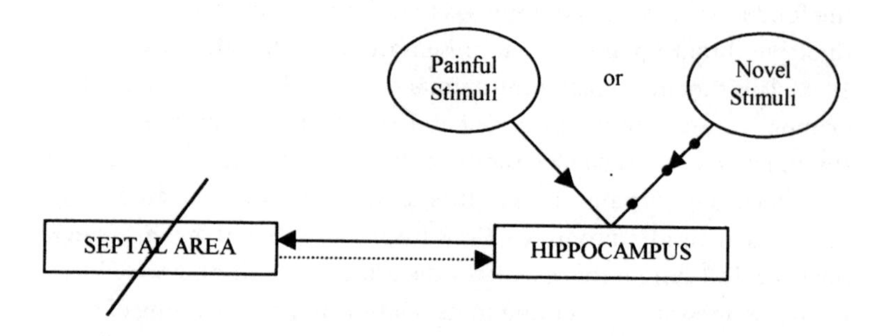

Fig. 8.1 Effects of septal lesion on hippocampal theta wave. (Based on data in Gray, 1970.)

mechanisms. If there is a discordance between actual and expected stimuli (novelty input to the BIS) or if the predicted stimulus is aversive ('signals of punishment' or 'signals of non-reward' to the BIS), the septo-hippocampal system takes direct control over behaviour and operates in the 'control mode'.

In control mode, the septo-hippocampal system operates the output of the BIS with the participation of other brain structures. In this way, the septo-hippocampal system operates as the computational mechanism of the BIS, and detects anxiety-producing stimuli, thus activating appropriate behavioural patterns. Although the functions attributed to the septo-hippocampal system are cognitive, the system plays a role in emotional behaviour. Gray's (1987) model treats this system as an interface between emotional and cognitive processes.

Experiments on the BIS

Given the role of the BIS in the reception of pain and novelty, it is not surprising to find that lesions to both the hippocampal and septal regions impair an animal's ability to withhold a response that had been punished. Such lesions also affect the animal's reactions in the non-reward situation. They maintain a response in the face of non-reward more so than non-operated control animals. Gray (1971) has shown that lesions in the medial septal area are the most effective in modifying the rat's reactions to non-reward and punishment.

Gray (1971) hypothesised that injections of the barbiturate, sodium amobarbital, and septal–hippocampal lesions produce similar behavioural effects because the septal pacemaker cells are inhibited by barbiturate drugs. Thus the theta wave from the hippocampus is abolished either by disruptions in septal functioning or by direct intervention of the hippocampal area. This hypothesis led Gray & Ball (1970) to implant recording electrodes in the dorsal hippocampus and stimulating electrodes in the medial septal area. By stimulating the septum at the theta range, the experiments were able to induce or drive a 6–9 Hz wave in the hippocampus. They also found that the theta wave could be blocked by stimulating the septal area with a current of higher frequencies, in the range of 100–200 Hz.

Once the researchers were able to induce hippocampal theta waves by septal stimulation, they experimented with the threshold level of stimulation that would produce this effect. The threshold current to drive theta is related to the stimulation frequency, and a minimal threshold was found at precisely 7.7 Hz. Anti-anxiety drugs increase the driving threshold selectively at 7.7 Hz, suggesting a relationship between theta at this frequency and anxiety. Such drugs have little effect on other frequencies in the theta range. Because recordings from the hippocampus reveal theta waves in the 7.5–8.5 Hz range when rats are exposed to frustrative non-reward, it is not surprising that such drugs attenuate their behavioural approach to cues associated with this event.

Gray (1987) proposed that 7.7 Hz theta driving mimics the central effects of conditioned frustration on the basis of behavioural effects affected by this physiological manipulation. Gray (1970) trained rats in a runway on a CRF schedule and divided them into various groups. The rats were tested under non-reward, and those in the experimental group were subjected to continuous theta driving while in the runway. Another group was stimulated in the septal area but at frequencies different from those that induce the theta rhythm. A third group did not receive any stimulation at all during the test trials. The results showed that the theta-driven rats showed a potentiation of the extinction process; that is, they extinguished sooner than the other groups. The other two groups did not differ in their rate of extinction but clearly were more resistant to extinction than the experimental group.

How anti-anxiety drugs are able to prevent animals from exhibiting the PREE as well as the partial punishment effect (PPE) has been discussed. Since these drugs also affect the 7.7 Hz theta-driving threshold, the latter effect may play a causal role in explaining PREE and PPE. To test this hypothesis, Gray (1971) 'injected' 7.7 Hz theta into the rat's brain through appropriate electrical stimulation of the septal pacemaker cells that control theta. By doing so he hoped to produce emotional 'toughening up' or behavioural tolerance to aversive stimuli through a sequence of theta-driving prior to test trials where the animal encounters aversive stimuli. Gray trained rats to run in a straight alley on a CRF schdule. The experimental group was given five seconds of theta-driving when they encountered the goal box. This was done randomly on 50 per cent of the trials. The control group was not stimulated. During extinction tests, both groups were left unstimulated, and the results indicated that the experimental group was more resistant to extinction than the controls. This outcome may be referred to as a *pseudo-PREE*, as shown in Fig. 8.2.

Holt & Gray (1983) extended Gray's (1971) findings by testing the animals on a different apparatus. The experimental group received 10 days of theta-driving stimulation and the control group was given identical training save for the stimulation. After this, the rats were trained to bar-press for food reward without further stimulation. The experimenters then attempted to suppress bar-pressing by extinction, punishment with foot-shock, or presentation of a stimulus paired with a response-independent foot-shock. In each of these tests the previously stimulated rats were more resistant to suppression of bar-pressing than the controls, even though the behavioural tests were conducted three to four weeks after the end of the period of brain stimulation. The fact that 'emotional toughening up' was produced through theta-driving applied prior to behavioural training suggests that the process which gives rise to PREE and PRP is non-associative.

It is possible to block hippocampal theta by high-frequency septal stimulation. If the critical aspect of low-frequency stimulation is that it drives theta, then high frequency septal stimulation should increase later tolerance for stress. Holt & Gray (1983) found that high-frequency, *theta-blocking* septal stimulation decreased the resistance of

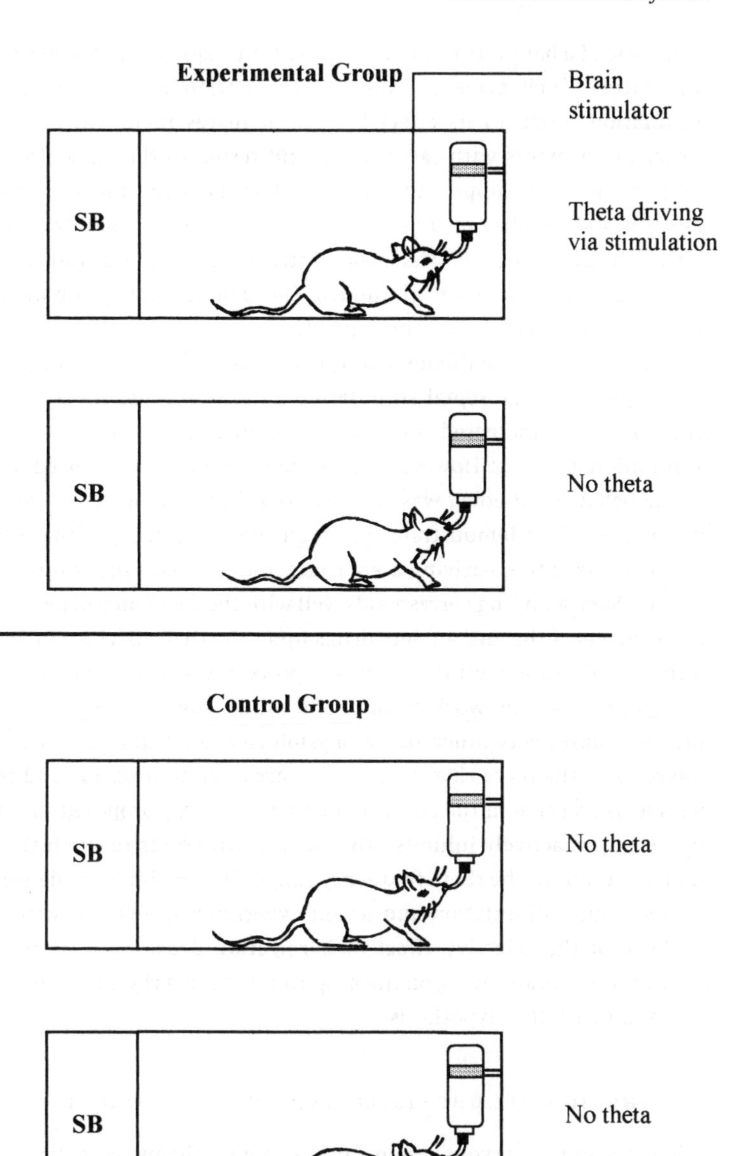

Fig. 8.2 The 'pseudo-partial reinforcement extinction effect' induced by theta-driving the experimental animals on 50 per cent of their continuous reinforcement (CRF) trials. (Based on data in Gray, 1969.)

extinction of a bar-press response acquired subsequently to the period of stimulation. Such evidence supports the notion that low-frequency stimulation produces its effect because it drives theta. Gray's (1969) earlier experiments with rats trained and tested in the alley also produced results that support this notion. The rats were trained on a PRF schedule and received medial septal stimulation with a 200 Hz current during the non-rewarded trials. The control group was also trained on a PRF schedule but was never stimulated. The experimental group was less resistant to extinction than the controls.

More recently, Williams & Gray (1996) have investigated the role of the frequency of the septal stimulating current and the theta rhythm which it elicits. Increased resistance to extinction was produced after stimulation at 7.7 Hz. However, when the frequency was reduced to 7.5 Hz the behavioural effect was reversed, that is, resistance to extinction was reduced. This demonstrates the exquisite dependency of the behavioural effects of theta-driving upon exact frequencies of the stimulating current. Such a finding corresponds well with the frequency-dependence of the effects of the anti-anxiety drugs upon the theta-driving curve. All of these findings suggest that common processes may be involved.

In general, the work by Gray and others suggests that theta frequency behaves very much like a physiological analogue of frustrative non-reward. The theta rhythm acts as a carrier of information and thus plays a critical role in the control of behaviour. Gray proposed that the hippocampus actively inhibits other neural structures through the repetitive output in the theta frequency range. The combined evidence on the recording, stimulation and lesion experiments, together with the evidence of the selective effect of barbiturate drugs on a particular frequency of septal-driving on the hippocampal theta rhythm seem to be consistent with this hypothesis.

THE FIGHT–FLIGHT SYSTEM AND ATTACK BEHAVIOUR

In his conceptual nervous system concerning mechanisms dealing with aversive reactions, Gray (1987) made a distinction between those that react to unconditioned aversive stimuli and those that react to conditioned aversive stimuli. The latter forms the BIS, and the former constitutes the *fight–flight system*. When an animal is exposed to a punishing aversive stimulus, it displays unconditioned behaviour. It will either fight or show flight. He also distinguished between two kinds of fight: the *defensive attack*, which is delivered against a member of its own species or a predator; and a *predatory attack* delivered by a predator against its prey. Such a distinction is supported by the type of aggressive behaviour elicited by electrical stimulation of the brain. Defensive attack is elicited by electrical stimulation of the medial hypothalamus in both cat and rat. In cats this behaviour is accompanied by hissing and erection of the hair (Masserman, 1941). This stimulation also produces unconditioned escape behaviour of the sort elicited by painful stimuli (Roberts, 1958). The unconditioned response to punishment may be either defensive attack or attempts at escape.

Adams & Flynn (1966) trained cats to escape from shock by jumping onto a stool. When these animals were stimulated in the medial hypothalamus, they also exhibited the same escape response. Using the self-stimulation method for mapping the brain into zones related to reward value, Olds & Olds (1965) found that medial hypothalamic stimulation can act as a negative reinforcer of the rats' operant responding. These animals were subjected to electrical stimulation of that area, and the equipment was programmed so that each bar press would terminate the stimulation for a specified period. Under this response-reinforcement contingency, the animals responded at a very high rate. Thus defensive attack behaviour may be motivated by activation of the negative reinforcement zones in the brain. In addition, there may be another factor that is reinforcing such behaviour. Rats exposed to an inescapable electric shock when placed in pairs exhibit a reduced physiological response (i.e. gastric erosions, as well as ACTH release) than those shocked when alone (Stolk *et al.*, 1974; Weinberg, Erskine & Levine, 1980; Weiss *et al.*, 1976). The Weiss *et al.* (1976) study also demonstrated that assuming the fighting posture without having contact with the partner separated by a Plexiglas barrier can also reduce the aversive effects of shock. They suggest that fighting is a coping response because it provides relevant feedback in the form of self-reinforcing behaviour. In this respect, Weinberg *et al.* (1980) consider fighting to be a consummatory response similar to others such as eating, drinking and mating, in which the performance of the act can be reinforcing in itself, quite apart from the incentive properties of the goal object (Valenstein, Cox & Kakolewski, 1970).

It is striking that the stimulation of the same neural region can elicit either defensive attack or escape behaviour. Whether fight or flight occurs is independent of the peripheral shock delivered or the neural region stimulated but on other stimuli present at the time of punishment (Gray 1987). A single fight–flight system receives information about all aversive events. Whether escape or attack behaviour occurs depends upon the total stimulus context in which the electric shock or brain stimulation is received. This notion is supported by the results of an experiment by Azrin, Hutchison, & McLaughlin (1965) in which monkeys were able to pull a chain to present themselves with a ball. When they were shocked the monkeys pulled at the chain at a high rate and bit the ball. However, if the experimenter provided the animals with the opportunity to perform the escape response that terminated the shock, the monkey rapidly came to prefer that reponse than to attack the ball. If the escape response was made ineffective again, the monkey returned to the attack behaviour. Such results suggest that the expression of fight or flight responses depends on the particular stimuli present in the animal's environment when punishment is delivered. This implies the operation of a single fight–flight mechanism which receives information about all punishments and then issues commands either for fight or for flight depending on the environmental context.

Predatory attack behaviour and positive reward

The preceding discussion dealt with mechanisms of defensive attack and did not consider attack behaviour of an appetitive nature. Studies on the cat indicate that stimulation of the lateral hypothalamus (LH) produces a predatory attack on a rat in which it silently stalks its prey, with hair sleeked back and no sign of fear (Wasman & Flynn, 1962). Self-stimulation of this region causes animals to show a very high rate of responding and suggests that it has positive reinforcing effects. Attack behaviour which produces such stimulation is closely connected with the act of eating or drinking, since these acts can be elicited by stimulation of the LH. Discussion of the role of LH in feeding and drinking is presented in Chapters 4 and 6, respectively. One can argue that predatory attack is essentially approach behaviour of the same kind as food-seeking or water-seeking. Roberts & Kiess (1964) found that if the LH stimulation was turned on when cats were eating, they switched their activity to attacking a nearby rat. This led them to conclude that attack behaviour of a predatory nature is positively rewarding. This contrasts with the effects of stimulation of the negatively reinforcing areas of the brain on the evocation of *defensive aggression.*

The overlap in the anatomical sites regulating predatory attack and other consummatory activities such as feeding and drinking is extensive. Valenstein (1973) found that stimulation of such sites also elicits gnawing as well as exploratory behaviour in rats. In addition, Caggiula (1970) and Stephan, Valenstein & Zucker (1971) have shown that the elicitation of male copulatory behaviour and feeding can obtained from the same electrode. The same electrode site can be used to stimulate consummatory activities as diverse as carrying of objects (Phillips *et al.*, 1969), grooming and digging (Valenstein *et al.*, 1970).

Electrical stimulation of the fastigial nucleus of the cerebellum also produces multiple effects such as alertness, cardio-vascular reactions, grooming, feeding and attack behaviours in the cat (Reis, Doba & Nathan, 1973). The intensity of the electrical stimulation and the availability of the goal object rather than the location of the electrode influenced the identity of the behaviour. For example, if a rat and food were both available, the cat shows attack behaviour, but in the absence of a prey, the cat would eat. These results once again demonstrate that the nature of the behaviour evoked from a single electrode at a fixed stimulus intensity can be changed by altering the availability of goal objects. This supports Valenstein's (1973) views on the plasticity of the central neural organisation of some behaviours. However, it should be emphasised that predatory behaviour and defensive attack behaviour reflect different motivational, hence physiological, substrates. Although the physiological mechanisms underlying predatory attack have much in common with those concerning feeding and grooming, they have little in common with the physiological basis of defensive attack. Perhaps these substrates are organised along the lines of appetitive and aversive events.

AVERSIVE STIMULI AND MOTIVATION SYSTEMS IN HUMANS

Like non-human animals who are exposed to aversive stimuli, humans also respond to such events by removing themselves from the source of unpleasantness or by attacking the source if the first option is made difficult. Exposure to noxious stimuli elicits stress responses such as muscle tension, hypertension and heart rate increase in humans as they do in animals. Cues in the setting that are associated with these sources elicit a state of fear or anxiety which are comprised of the set of internal stimuli associated with stress. Gray (1987, p. 204) defined anxiety as 'that emotional state which is elicited by stimuli associated with either punishment or non-reward'. These kinds of stimuli are termed 'anxiogenic' and drugs that reduce their impact are descibed as 'anxiolytic'. Human modes of emotional reaction may be essentially similar to those of animals. Anxiolytic drugs such as alcohol and barbiturates affect the behaviour in animals as they do in humans. Gray regarded the parts of the brain that mediate responses to anxiogenic stimuli as belonging to the BIS discussed in the previous section.

There is certainly widespread use of alcohol, barbiturate and anxiolytic drugs such as benzodiazepines among people faced with difficulties. Such difficulties rarely involve the sort of punishing stimuli that cause 'fear' but instead, involve frustrated hopes and/or unfulfilled expectations. Gray (1987) considered the septo-hippocampal system the first place to examine when studying differences between people with high or low levels of susceptibility to anxiety because these structures form the corpus of BIS functioning. This was based on the study of Reiman *et al.* (1984) using positron-emission tomography (PET) scanning as a method of visualising activity levels in various brain regions. Patients suffering from spontaneous attacks of panic differed from controls in the inputs and outputs of the septo-hippocampal system.

Although benzodiazapine and related drugs are used to treat anxiety disorders, they act only to suppress these reactions temporarily. Gray (1987) saw more hope with behaviour therapy treatment because of the many studies which he reviewed indicating how rats increased their resistance to frustrative non-reward through PRF training, as to punishment through the partial punishment schedule. Such contingencies of exposure to anxiogenic events lead many people to toughen up emotionally.

Sources of human aggression

The most widely cited classification of sources of aggression is that of Moyer (1977, 1980) who distinguished among different systems of aggressive behaviour on the basis of their arousing stimuli. One of his classifications, irritable aggression, concerns behaviour that is aroused by a range of aversive conditions, such as painful ones. He also proposed *fear-induced aggression* as another category. These forms of aggression are comparable to those of defensive aggression discussed earlier. They are documented by findings indicating that pairs of rats can be induced to fight by painful electric shocks and that a monkey that receives a shock may learn to pull on a chain to produce an object that can then be attacked (Azrin *et al.*, 1965). Comparable studies on humans involve someone insulting a subject, or otherwise treating him or her badly. A little later, the subject that has a chance to deliver electric shocks to the insulting person, is more likely to deliver shocks than subjects who had not been insulted. It should be noted that in such experiments, the subject only thinks he or she is delivering shocks, whereas in fact no shocks occur (Berkowitz, 1983, 1989). The likely reason for such agonistic behaviour is the activation of a *defensive motivation system* akin to that encountered when we examined the effects of the medial hypothalamus stimulation on rats.

Hunting animals is a popular activity among many humans in the past as well as the present. In modern times such activity ranges from the fox-hunt in England, wild game hunting in North America, to African safaris. Despite protests from animal rights activists, hunting is an activity actively sought by many males who live in an urban setting. Wallace (1979) described hunters from Los Angeles who pay dearly for the privilege of being flown in with their high-powered rifles to Santa Cruz Island to kill the hapless descendents of the domestic sheep that were once raised on the island. The killing cannot be for the stringy meat of these animals; it seems to be done for the pleasure of killing. In analysing the evolution of hunting, Washburn & Lancaster (1968) stated that 'men enjoy hunting and killing'. The opportunity to hunt wild animals serves as an incentive for instrumental activity. Hunters exhibit instrumental activity which precedes the actual consummation of the predatory kill. There appears to be reinforcement or reward from the accomplishment of such an act. Is this an expression of the motive of *predatory aggression* which we discussed in the section on physiological mechanisms in animals?

Predatory attack behaviour is usually regarded as a class of behaviour in which the animals that spontaneously kill, such as cats, are influenced by a complex motivational system which is independent of, although linked to, alimentary and aggressive mechanisms (Zagrodzka & Fonberg, 1978). It also involves a strong hedonistic component in the sense that there appears to be 'enjoyment' or 'pleasure' experienced in the act. Although one might view predatory attack in carnivores as a means to secure food, Zagrodzka & Fonberg's (1997) summary of their laboratory findings indicate that food motivation is not necessary to drive such behaviour. These researchers proposed that some forms of human aggression may be understood from the perspective of the animal model. Predatory behaviour, in the sense of chasing and finally killing the prey, is rewarding to a cat, and might be functionally autonomous from the feeding motivation. In this respect, predatory behaviour resembles human intrinsic aggression in which the 'organism's satisfaction appears to be intrinsic to the aggresive act itself' (Feshbach & Fraczek, 1979). The reinforcement may be associated with the infliction of pain or with the performance of an aggressive response. Fraczek (1992) has also argued 'aggression for pleasure' is based on reward mechanisms.

Homicide

Homicide (humans killing other humans) is a mystery. Intraspecies aggression occur in other animals, but normally it seldom gets carried to the 'bitter end'. The classic analysis by ethologists such as Lorenz (1971) and Tinbergen (1968) explains this anomaly of extreme intraspecies aggression as follows. Animals like humans compete for resources and status, for the reasons elaborated in Chapter 2. When the outcome of these agonistic encounters is established (there is a 'winner' and a 'loser'), there seem to be behavioural reactions by the winner triggered by built-in signals that are innately displayed by the loser. The latter displays submissive gestures that serve to inhibit the release of aggressive behaviour by the former. For example, in dogs and wolves, the defeated animal assumes a posture which exposes its neck to the winner. Similarly, a jackdaw abandons his attack and preens the intended victim when the latter turns his head, thus exposing his silver nape feathers. In both cases, attack by the victor is inhibited by these stimuli. Such mechanisms are lacking in humans.

The more 'modern' evolutionary psychology approach to homicide is exemplified by that of Daly & Wilson (1988, 1994) whose position is that the act of murder is *not* an adaptation, but a manifestation of the human mind that has been adapted in such a way that, under certain circumstances, murder is a likely outcome. It is not the behaviour itself but the psychological mechanisms that bring it about which is of interest to these researchers. There are some consistent patterns of gender and age and motive that describe the homicidal population. *Homo sapiens* is clearly a creature for whom differential social status has been associated with variations in reproductive success. In traditional societies, men of high rank had more wives, concubines and access to other men's wives than men of lower social rank. As a result such men had more children and their children survived better. As discussed in Chapter 2, the situation in modern industrialised societies is that there is an inverse relationship between class and status and number of children produced. Perusse (1993) suggested that the greater use of contraceptives among the more educated higher status males is one of the reasons for the modern situation, and had there not been this invention, the situation would be similar to that in traditional societies. The proximate causal factor that is driving all this is 'pleasure', and high status professional men (at least among his francophone sample in Montreal) seem to have more opportunities for sexual encounters than the less affluent. So what has been selected is *not* the conscious desire to leave behind more children but the propensity to engage sexual activity for its own sake as well as the status associated with it among men

In early *Homo sapiens*, individuals' competitive success or failure would have enhanced their reputation among each other, and this could affect their total lifetime survival and reproductive success (Daly & Wilson, 1994). Demonstrations of bravery in the face of danger and before an audience enhances a male's reputation. One of the motives of modern male–male homicide is the defence of status, reputation and honour in the local peer group. A seemingly minor affront can enrage young men to engage in violent encounters. Daly & Wilson (1988) claim that in most milieus, a man's reputation depends upon the maintenance of a credible threat of violence. They cite 1975 homicide statistics in the United States that indicate that youth from puberty to the mid-twenties constitute the group that is most victimised and also the instigator of violence. Daly & Wilson (1994) regarded this demographic class as the group upon which there was the most intense selection for confrontational competitive capabilities among our ancestors. The group committing the most murders are young men engaged in trivial altercations and some getting killed as a result.

The next most frequent basis of homicide is within the family. If the implication is that murderers kill their kin, kin-selection theory is in trouble. But when Daly & Wilson (1988) examined the reports from American and Canadian sources, it turns out that most 'family' victims were the murderer's spouse. Sexual jealousy seems to be the motive that triggers such aggression and homicide. Men who kill their partners typically do so under two key conditions – the observation or suspicion of sexual infidelity or when the woman is terminating a relationship. The first places a husband with a family at risk of investing his limited resources in an offspring to whom he may not be genetically related. The second represents the loss of a productively valuable mate to a rival, which represents a loss in the currency of fitness. Although men do not carry such evolutionary logic in their heads, they do carry the psychological mechanisms that led to their ancestors' success. This can fuel sexual jealousy and proprietariness over mates, both of which can lead to aggression. The results of interviews with 277 women who had been assaulted by their husbands led support to the hypothesis that violence by men is used as a strategy for controlling their mates, with the goal of preventing sexual access to other men or a defection in their relationship (Wilson, Johnson & Daly, 1995). Violence against women is dependent upon the specific adaptive problem being faced by the man.

SUMMARY

Aversive stimuli elicit adaptive reactions such as withdrawal or attack directed at the source of the stimuli. Animals acquire a conditioned response, namely fear, to cues or warning signals associated with aversive stimuli, and also learn successful avoidance responses that reduces the fear. A more evolutionary based explanation suggests that successful escape or avoidance responding does not strengthen a new response, but that the animal is programmed to make specific responses when threatened with danger. What is learned is that certain situations are dangerous, and that a specific member of a class of *species-specific defensive reactions* is effective.

Although some stimuli are more closely associated with fear than attack, and vice versa, a range of stimuli may evoke either reaction. Much depends upon the environmental context, and the animal's previous encounters with such stimuli. In addition, the occurrence of non-reward in situations where animals have experienced reward also has aversive effects. Gray's (1987) model indicates parallels between the effects of punishment and those of the omission of anticipated reward in their behavioural as well as physiological manifestations. This model also distinguishes between mechanisms dealing with unconditioned aversive reactions (the *fight–flight system*) and those with conditioned aversive stimuli (the BIS).

9

Social motivation: attachment and altruism

ATTACHMENT ACTIVITIES

Motivational influences that are specifically social constitute the subject of this chapter. Social motives involve activities that affect interactions among organisms of the same species as well as organisms of other species. I will deal with only two issues in this short chapter: the formation and maintenance of social bonds between individuals; and that of pro-social and altruistic acts. The functional and proximate causal aspects of these phenomena will be considered in the analysis. The first and foremost social motive concerns the formation of the bond between an infant and its primary caretaker, usually the mother. The material in Chapter 3 deals with this issue from the perspective of the mother, whilst the material in this chapter complements it through an analysis of processes in the infant. The attachment of a human infant to its caretaker has biological roots. Both babies and adults are programmed by evolution to become attached in certain ways because the former is dependent upon the latter for survival. Attachment behaviours refer to a broad classification of behaviours that keep the infant in close proximity to an attachment figure. These behaviours include crying, clinging and approaching, as well as others produced when the infant is separated from the attachment figure.

Mammals live in a diverse array of habitats and social structures. The basic unit of the family is the mother and infant. However, our examination of material in Chapter 3 suggests that other conspecifics may be involved to some exent in infant caregiving. These caregivers can include the father, siblings or peers, or combinations of these conspecifics. When mothers are unavailable, father–infant contacts may increase, and when fathers are involved, they generally direct their care to kin rather than to non-kin. Infants generally display attachment responses to individuals that provide early nurturing contacts with the infant. Whole sensorimotor organisation is involved, and implicate at least two features. The first feature which is discussed from the perspective of the mother in Chapter 3, involves recognition and discrimination based on one or more sensory modalities. The second feature involves the more complex integrative cognitive system that the animal uses to predict and relate to the world. This internal model is particularly evident amongst primate species, especially humans.

Most species display atttachment behaviours that in a normal environment contribute to the survival and inclusive fitness of the individual. Primate species are more likely to show individual recognition and preference than non-primates because of differences in the cognitive structures or internal working models amongst the former. This ability amongst primates may be related to the highly developed discriminative functions necessary for individual recognition systems that are regulated by neocortical development in the primate brain (Crnic, Reite & Shucard, 1982; Reite & Boccia, 1994). Most sensory organs necessary for attachment in primates are functional at birth or shortly thereafter, and continue to develop and change. Primates are generally altricial, and this results in a long period of nursing. Females generally deliver only one infant and assume primary care for it. Sometimes the father, siblings and other conspecifics may participate in infant care. Because of the increased time required for brain development, primates generally have a longer period of gestation and dependency during infancy. Most primates carry their infants, and infants cling to them, all of which contributes to the formation of primate attachments.

There is a sensitive period in which attachment is most likely. The more precocial the infant, the shorter and more rigid the sensitive period. The more altricial the infant, the longer and more flexible the sensitive period. Rhesus monkeys isolated for the first three months of life can develop essentially normal behaviour when placed in a nurturing social environment (Harlow & Mears, 1979). If the period of isolation is past the time when socialisation normally begins, then attachment is difficult, as well as recovery. In humans, the development of attachments follows four stages according to Bowlby's (1969) system. First, they show indiscriminate responsiveness (0 to 2 months), then discriminant responsiveness (3 to 6 months) involving increasingly caretaker preference, then attachment proper (6 months to 3 years), as evidenced by distress if the caretaker goes away and fear of strangers, and finally, a change in their attachment relationship to a 'partnership', as well as a generalisation of the attachment bonds to others.

Bowlby's (1969) theory proposes that attachment is characterised by four components. (1) *Instinctive infant attachment behaviour*. The mother has stimulus characteristics that elicit the behaviour of the infant. In non-human primates, initial attachment behaviours include grasping and clinging. In humans, signalling behaviour such as crying, babbling and smiling as well as visual tracking predominate until the infant's motor systems mature. (2) *Goal-directed behaviour*. Attachments functions as a goal corrected system. The primary goal is to maintain a perception of security within the infant and to attain conditions that produce it. Goal-corrected behaviour leads to the development of a specific attachment as a result of: an in-built preference towards looking at certain patterns in preference to others, 'exposure learning' by which the familiar becomes distinguished from the strange, an in-built bias towards approaching the familiar, and feedback of results which augments a behavioural sequence. (3) *Internal representations and working models*. The infant develops dynamic cognitive working models of the mother and the surrounding world. It uses these models to predict the behaviour of the mother and thereby to regulate its own behaviour. (4) *The secure base*. The presence of the mother facilitates exploratory behaviour. When no apparent danger threatens, the child can explore at a fair distance from the mother, but when stress-arousing stimuli are present, the attachment system pulls the child closer to available protection.

Freud & Riviere (1930) regarded social attachment as biologically adaptive and as a means through which society introduces a controlling overlay of more basic behavioural systems such as those that regulate sexual behaviour, food and water intake, fear responses, aggression and self defence. Using concepts developed in ethology, Bowlby (1969) proposed that the baby is a powerful source of releasing stimuli that affect adults. These elicit caretaking responses from adults which the baby requires. Parents are seen as responding not to the baby's needs but to its signals. Early interactions between the child and caregiver form a back-and-forth chain of stimuli and reactions to them. Out of this grows a strong emotional bond between infant and caregiver, namely attachment. Because the basic behaviour regulatory systems of the infant are acquired through interactions with a caregiver, the process is more like a dance than a formal transfer of knowledge. For example, if a human baby smiles at an adult, the adult is likely to smile back, and if she or he does not, the baby may begin to cry. A baby's cry is a very powerful stimulus for adult action because it elicits increased blood pressure and expressions of stress even in adults who are not parents (Frodi & Lamb, 1978).

PROSOCIAL AND 'ALTRUISTIC' ACTIVITIES

Altruism is defined as behaviour that incurs a cost to an individual but which provides a benefit to another (Hamilton, 1964a,b). For altruism to evolve, the cost to the donor must be less than the benefits provided to the recipient, multiplied by the genetic relatedness between the donor and the recipient. Selection will favour adaptations for helping kin in proportion to their genetic relatedness. Hamilton predicted that selection will favour the evolution of mechanisms to help close kin more than distant kin and distant kin more than strangers. The favouritism that parents show their own children can be viewed as a special case of favouritism toward the 'vehicles' that contain copies of their genes.

Alarm calling in Belding ground squirrels

When these ground squirrels detect a predator such as a badger or coyote, they sometimes emit a high-pitched staccato whistle that functions as an alarm call which alerts other squirrels in the immediate vicinity. The alerted squirrels then run for safety and avoid being caught by the predator. The alerted squirrels clearly benefit from the alarm call because it increases their odds of survival. However, the alarm caller suffers because the whistle makes it more easily detectable by the predator. This act of altruism where the donor increases the fitness of the recipient at some cost to itself, seems contrary to individual survival. Sherman (1977, 1985) spent summers in the California woods studying these animals by marking, tracking and studying an entire colony, and found the following. Sounding the alarm puts the signaller at risk because predators were observed stalking and killing alarm callers at a higher rate than non-calling squirrels in the vicinity. He noticed that females give alarm calls more often than males, and that females remain within their natal ground while males leave home and join non-related groups. Thus females are surrounded by sisters, daughters, aunts, nieces and other female relatives. This finding supports the inclusive fitness hypothesis (Hamilton, 1964a,b) that an individual's relatives are all vehicles of fitness though differing in value.

The ground squirrel data support the idea that altruism can evolve through the process of inclusive fitness. Females who do not have daughters or other children in their vicinity but who have other genetic relatives such as sisters, nieces and aunts close by emit the alarm call when they spot predators. In another study, Holmes & Sherman (1982) found that female ground squirrels assist their sisters and daughters in territorial disputes with invaders, but do not help non-relatives in such conflicts. Although the evolution of altruism among genetic relatives has been explained, one wonders how it could evolve among non-relatives, given the selfish designs that tend to be produced by natural selection. This is what evolutionary biologists call the problem of altruism.

Social exchange and reciprocal altruism

There is evidence that a form of co-operation known as social exchange occurs amongst humans as well as other primates. The theory of reciprocal altruism states that the psychological mechanism for providing benefits to non-relatives can evolve as long as the delivery of such benefits is reciprocated at some point in the future (Axelrod & Hamilton, 1981; Williams, 1966). Reciprocal altruism can be defined as co-operation between two or more individudals for mutual benefit (Cosmides & Tooby, 1992). In evolutionary terms, those who engage in reciprocal altruism will tend to outreproduce those who act selfishly. This would cause psychological mechanisms for reciprocal altruism to spread in succeeding generations.

Reciprocity in vampire bats

Vampire bats depend on the blood of other animals as a source of food. They live in groups of females and associated offspring. The males leave the colony when they are capable of independence, a characteristic also seen in the Belding ground squirrels. Vampire bats hide during the day, but at night, they emerge to suck the blood of cattle and horses. Their victims are not exactly willing donors. Because the bats' ability to feed successfully increases with age and experience, 77 per cent of those under two years old get blood on any particular evening, whereas 93 per cent of the older bats are successful (Wilkinson, 1984). The problem of younger bats raises the question of how they survive failed attempts to obtain food. Failure at feeding can quickly lead to death, as bats can not go without food for more than three days. Wilkinson discovered that the bats regularly regurgitate a portion of the blood they have sucked and give it to others in the bat colony. This aid-giving is not random but is given to those from whom they received blood in the past. The closer the association between the bats, and the more often they were sighted together, the more likely they were to give blood to one another.

Wilkinson (1984) also used a captive colony of vampire bats to explore other aspects of reciprocal altruism. He selected some bats for food deprivation and varied the length of time they were deprived. Wilson discovered that the friends tended to regurgitate blood more often when their friends were very close to starvation than when they were in mild need. He also found that the starved bats who received help from their friends were more likely to give blood to those who had helped them in their time of need. All of these behaviours suggest that vampire bats have evolved reciprocal altruism adaptations. Individuals are in frequent contact with one another over long stretches of time, and the frequency of association predicts the degree of altruism. The cost of regurgitation is relatively low when the bat is a donor but the benefit is high when it is a recipient. Unsuccessful feeds are common and equally likely to happen to any member, especially amongst the young ones, so the roles are likely to alternate. Individuals can recognise one another, and the likelihood of reciprocal altruism is high.

Reciprocity among baboons and vervet monkeys

Savanna baboons are intensely social animals that interact frequently in groups. They often compete over access to limited resources such as food or sex partners, and conflict erupts on a regular basis. In the course of studying a troop of baboons over a lengthy period, Packer (1977) witnessed many requests for help in the course of such social conflicts. During a conflict, baboons will solicit help by establishing eye contact with the baboon from whom they seek help and shift eye contact between the respective helper and the antagonist. In 20 cases of this sort, the solicitation involved a conflict between males over sexual access to a female baboon who had entered oestrous. Requests for help from a third male baboon in these sexual conflicts tended to elicit aid more often than requests for aid in other contexts. The request for aid yielded success in 16 of the 20 encounters, and the female then mated with the baboon who had previously asked for help. In this particular situation the helper never benefited directly by giving aid because he never received direct sexual access as a consequence of his help. The ability of baboons to recognise one another and to remember the outcome of their previous interactions facilitates the evolution of reciprocity. This pattern of reciprocal altruism is so strong amongst baboons that males who fail to form friendships of this type rarely have sexual access to females.

Other researchers have not been able to observe the reciprocity seen by Packer (1977). Noë, de Waal & van Hooff (1980) found that when the fifth- to seventh-ranked males in a band formed coalitions of two or three males, who separated a high-ranking male from a female, the fifth-ranked male won the female in 17 of 18 cases. Reciprocity was not evident in these males' interactions. Noë *et al.* (1980) suggested that if lower ranking males have no chance of mating on their own, then the chance of an occasional mating by themselves may be sufficient to repay them for joining coalitions, despite the absence of reciprocity. Bercovitch (1988) studied olive baboons in Kenya and found that coalitions are generally formed by older, middle-to-lower ranking males and targeted at younger, higher-ranking males. However, males who solicited others for coalition formation did not necesarily gain access to females more often than their fellow coalition members.

Reciprocal altruism was observed, however, among the vervet monkeys in East Africa. Seyfarth and Cheney (1984) conducted field experiments on the effects of grooming on the tendency of vervet monkeys to ask for help at a later time. The experimenters first played tape recordings of females emitting vocal solicitations for help. They then played back these tape recordings to other females who had been groomed recently by the solicitor, and found that females who had been groomed recently by the solicitor were more likely to look around in response the request for aid. Those who had not been groomed recently ignored the request.

Alliances are central features in the social lives of chimpanzees (Waal, 1982). Males regularly solicit alliances with females, grooming them and playing their infants. Without alliances with the females, males could never attain dominant status in the troop. If a female is found associating with an opponent, a male will bite or chase her. But shortly therafter, when she is no longer associating with the opponent, this male will be extremely friendly, showing great solicitude toward her and her infants. This appears to be a major strategy in the formation of chimpanzee alliances, that of severing the alliances of one's opponents and to enlist them again. Males defend the females from attack from other males and act as peacemakers in disputes. In return, the females support the males, aiding in the maintenance of their status.

Altruism in humans

There are many examples amongst humans where an individual engages in behaviours with a possible fitness cost to the donor whilst providing a fitness benefit to the recipient. Evolutionary analysis suggests genetic relatedness is an important determinant of acts of helping and altruism.

Genetic relatives are the vehicles for our fitness, and the closer the relative, the more beneficial an act of helping will be to our fitness. Selection pressures over human evolutionary history would have likely resulted in psychological mechanisms that lead us to aid close relatives more than distant ones. Burnstein, Crandall & Kitayama (1994) studied subjects from American and Japanese cultures by asking them what they would do in two different types of scenario. One involved helping that is substantial, such as acts that affect whether the recipient will live or die, and the other involved helping that is relatively trivial, such as giving a homeless person some spare change, or picking up some small items from a store. Helping in these hypothetical scenarios decreased as the degree of genetic relatedness decreased. The age of the potential recipients also influenced the potential donor's decision. Helping in the life-and-death scenario declined as the potential recipient's age increased, whilst the responses were reversed in the trivial helping situation. The experimenters' prediction that helping an older relative would have less impact on one's fitness than helping a younger relative was confirmed by the results dealing with helping behaviour in hypothetical situations requiring more substantial commitment. Similar results were obtained with both cultures studied.

In the 'real world' of humans, past and present, there are countless incidences of altruistic actions by aid workers to strangers. There are organisations such as Médecins sans Frontières (MSF) or Doctors Without Borders, United Nations Relief Agencies, the Red Cross, the Red Crescent and Oxfam, among many others. Many of the workers in these organisations die or are held hostage, imprisoned, or undergo many risks of these sorts. They provide food and clean water, dig latrines, set up first aid stations and field hospitals, and often are the last to remain in situations abandoned by others as dangerous. In describing the work of MSF in Rwanda in 1996, Leyton & Locke (1998) state that in the crisis zone, MSFers act as if they do not care they are taking terrible risks. In explaining the motives for such actions, the authors suggest that working for MSF liberates them as human beings, freeing them from the trivialities of personal woes and the mindlessness of modern life. In acting thus, their personal dilemmas dissolve and their identities fuse. This explanation suggests a psychological basis of altruistic acts having to do with emotional connection and 'pleasure'. The authors' analysis implies that something akin to Maslow's (1954) 'self-actualisation' concept may be the motivating basis for the MSF workers' altruistic acts.

There are many other examples of volunteer self-sacrificial bravery found in warfare and in politically motivated terrorism settings so chaotic that institutional coercion is unlikely to be a factor. Terrorist organisations in the Middle East provide good examples of this. Campbell & Gatewood (1994) declared that present motivational theories, including those of sociobiology which focus on individual optimisation of inclusive fitness, seem inadequate to explain such behaviour. They suggested that that such self-sacrificial behaviour must have evolved from a setting in which loss of membership in one's group was more lethal than the costs of occasional heroism in hunting large animals and in intergroup warfare. Those early human groups must have been tightly bound, with limited membership, and no individual survival route through joining another group. Empirical support for this conjecture of ancestral conditions is difficult if not impossible to verify.

Another explanation suggests that altruistic actions are more likely to come from individuals who have 'status' within the community, and that there is selection for them. I will present this analysis as it was formulated to explain the motivation for risk-taking sentinel activity in a species of bird. Zahavi (1977) and Zahavi & Zahavi (1997) conducted field studies in Israel on the Arabian babbler, and noticed that individuals compete to serve as sentinels. They jockey amongst themselves for the 'honour' of guard duty, vying for the most dangerous post, the longest vigil or the hottest hour of the day. Is the bird doing this to help its kin or to reciprocate favours? Not so, according to Zahavi. He proposed that it is doing so to help itself in spite of the danger. In serving this role, the bird is communicating to others in the flock that it is strong, robust and alert enough to to bear the burden of sentinel duty, to take on the costs and still be able to thrive. Only an individual of high quality could afford to handicap itself so much. In Chapter 2 we surveyed data indicating that high status in the animal world is related to reproductive success. Zahavi (1995) argued that 'altruists' improve their fitness as a result of *honestly* signalling their phenotypic quality to prospective mates or allies. He suggested that co-operative systems may possibly evolve signals, using altruism as a handicap. For example, a female who chooses her mate on the basis courtship feeding will be more likely to have greater reproductive output than one that chooses a mate who displays less 'altruism'.

Humans who engage in relief and aid work dealing with helpless victims of war and famine, as well as workers engaged in other humane activities concerning the poor and underprivileged in our own societies are certainly regarded very highly. They often receive public acclaim, are awarded medals for bravery and humanitarian activities, and there is little doubt that they are regarded as high quality individuals. Although such individuals risk their lives, and in some cases, actually lose their lives in helping strangers, their names are remembered. Their family and relatives, in turn, are treated with great respect and honour. Altruism might act as a signal of quality, as proposed in Zahavi's (1995) handicap principle. Thus, it is possible that the mechanism of indirect selection leads to the enhancement of the individual's inclusive fitness.

I present a further example of altruistic actions of another heroic individual. The former chief hostage negotiator of the United Nations, Giandomenico Picco (1999) helped negotiate the release of more than a dozen western hostages held by Islamic militants in Lebanon by offering to meet the hostage takers personally under dangerous conditions. He willingly accepted being transported to various locations under blindfold by people whose identities were unknown to him. His life was at risk, but he did what he had to do because the lives of dozens of hostages were at greater risk. By winning the trust and respect of his captors, Picco managed to forge an unlikely coalition between Iran, Syria, Israel and the Lebanese groups to win the release of the captives. During the final stages of the negotiations, he offered himself as a hostage as a guarantee to the militants that the conditions of the agreement would be kept. In explaining why he risked his life in this and other diplomatic ventures, Picco wrote that he wanted to leave his son with this thought: 'principles matter more than life itself'. For Picco, his values and ideals matter more than the survival of the physical or biological structure which formulated these ideals or principles. This personal commitment may be explained from an evolutionary perspective if we consider the concept of *memes* (Dawkins, 1976).

All life evolves by the differential survival of replicating entitities such as the gene, or DNA molecule. Dawkins (1976) considered the possibility of other kinds of replicator and other, consequent, kinds of evolution, and in doing so, proposed the concept of *meme.* This term represents a unit of cultural inheritance which serves as the replicator of human culture. Just as genes propagate themselves in the gene pool from body to body via eggs and sperm, so do memes propagate themselves in the meme pool by moving from brain to brain via the process of imitation. Gene-selected evolution, by making brains, provides the structure from which the first memes arose. Once self-copying memes had arisen, their own evolution is continued. Dawkins (1982, p. 112) later conceded that 'memes are not strung out along linear chromosomes, and it is not clear that they occupy and compete for discrete "loci", or that they have identifiable alleles. The copying process is probably less precise than in the case of genes. Memes may partially blend with each other in a way that genes do not.'

The first rule of memes, as for genes, is that replication is not necessarily for the good of anything. Replicators that flourish are the ones that are good at replicating, for whatever reason. In past and recent history we have seen instances when a dramatic and well-publicised martyrdom inspires others for a deeply loved cause. This, in turn, inspires others to die. Although imitation is how memes can replicate, some memes are more successful in the meme-pool than others. Examples of memes include faith or blind trust in a principle, the notion of existence after death, the concept of hell for those who lived a 'bad life', patriotism, the notion of 'do unto others as they would do unto you', environmental awareness, arms reduction, abolition of slavery, political correctness, etc. Then there are pernicious memes such as hijacking airlines, 'copy cat killing', road rage and racial hatred.

When we die, we can leave behind us both our genes and our memes. The genetic aspect of us will be considerably diminished after three generations. As each generation passes, the contribution of your genes is halved, and it does not take long before their influence is neglible. However, if an individual contributes a 'good idea' to the culture, it may live on, intact, long after your genes have minimal representation. A cultural trait may have evolved simply because it is advantageous to itself. Dawkins (1976) acknowledged that a trait of humans, such as the capacity for genuine, disinterested altruism, may have evolved memically. Although humans like other organisms are fundamentally selfish, Dawkins proposed that, in addition, we have the capacity of conscious foresight. This is an ability to simulate the future in imagination, and provides us with the mental equipment to foster our long-term selfish interests rather than our short-term selfish interests. He also stated that although we are built as gene machines and cultured as meme machines, we have the power to turn against our creators.

Consider other examples of pro-social behaviour that are less heroic and dramatic than the infrequent ones that we have just considered. Most people dining in restaurants when on vacation leave tips for the server, even though they never expect to visit that place again. Many people will take a lot of trouble to dispose of unwanted pesticides, paints, antifreeze, used engine oil, old plaster wallboards, etc. by driving to a proper disposal depot often some distance from home rather than simply pour the liquids through the drain. The former activity is time-consuming, inconvenient, yet is practised by the majority of people. The probability of being detected and punished for doing something ecologically wrong is extremely low, and is an unlikely motivator of such socially approved behaviour. Besides the punishment is minimal, even if the individual is 'caught'. Such behaviour is very common in the Western world. Frank (1994) suggested that co-operation is motivated by emotions such as sympathy and guilt and that there are observable manifestations of these emotions that enable individuals to discern them in potential interacting partners. This would result in co-operators pairing with one another and avoiding interactions with defectors. If people with altruistic predispositions are observably different from others, altruists can interact selectively with one another and thereby reap the fruits of co-operation. He viewed altruism as a strategy that can be reproductively advantageous to individuals who follow it. In this manner, altruistic behaviour is analysed as a phenomenon favoured by the forces of natural selection.

SUMMARY

The bond between an infant and its primary caretaker, usually the mother, is the first and foremost social motive. Both infants and their mother are programmed by evolution to become attached. Attachment behaviours refer to a broad set of activities that keep the infant in close proximity to the attachment figure. There is a sensitive period in which attachment is most likely. The length of this period is related to the developmental rate of the species. Early interactions between the infant and the caregiver form a back-and-forth chain of stimuli and reactions to them. Out of this grows the emotional bond between them which characterises attachment.

Altruism involves behaviour that incurs a cost to an individual but which provides benefits to another. Natural selection favours adaptations for helping kin in proportion to genetic relatedness, and the individual's inclusive fitness is thereby optimised. This notion was supported by the results of studies on alarm calling animals. Reciprocal altruism which involves co-operation between non-related individuals for mutual benefit, occurs when individuals can recognise one another. Reciprocity has been documented in vampire bats through food-sharing, and social alliances among individuals in primates such as savanna baboons, vervet monkeys and chimpanzees. There are many examples of human bravery, self-sacrificial behaviour that cannot be explained in terms of optimisation of inclusive fitness. Humans engaged in aid work often risk their lives helping people of different backgrounds. Such personal commitment may reflect the the influence of *memes* or units of cultural inheritance that are self-replicating through the process of imitation (Dawkins, 1976). At the proximate level, aid-giving may be motivated by the experience of emotional connection as well as guilt and sympathy.

10

Conclusions and retrospective

MOTIVATION AND FITNESS

I trust that, after studying the material presented up to this point, the reader has a fair idea about the various factors underlying such motivated behavioural activities as mating, parental care, feeding, food selection, drinking, stimulus seeking and defensive or agonistic reactions. The material in the preceding chapters indicate that although the response patterns may differ, these motivated acts have one important feature in common. They are terminated when an end-point corresponding to a goal is achieved. The concept of fitness was discussed in the early part of this book with the presumption that organisms have evolved in such a way to maximise fitness. I discussed how evolutionary processes may have selected mechanisms that result in a particular behaviour. From this perspective, the functional significance of any aspect of behaviour is seen in terms of fitness maximisation or the chances it gives to the perpetuation of the organism's genes. The functional significance of mating behaviour is obvious, even though it may not directly enhance the survival of the individual. Although a motivated act such as feeding contributes directly to the survival chances of the individual, it does so through its contribution to the chances of gene perpetuation. The traditional distinction made between activities such as feeding, drinking or attacking and activities concerned with gene perpetuation such as sexual behaviour is unwarranted. All behaviour has been affected by evolutionary processes with consequences on fitness.

In discussing sexual selection, I presented material indicating that sexual dimorphism and sexual behaviour may decrease an individual's survival even though such attributes increase the chance of gene perpetuation. For example, males in mammalian species are forced to compete to gain access to a female or acquire and defend territory to which females are attracted. Fighting is a costly enterprise even for successful males because of injuries as well as eventual replacement when they age and are faced with a series of challenges. Courtship displays may attract predators, and some animals are made very vulnerable to predation while in the act of mating. Males, through the sexual selection of characters that attract the female, also have decreased chances of individual survival. There is a trade-off between motivational systems. In order to be successful in mating, males exhibit traits that make them successful in competing against each other, but in doing so, they incur a cost in survival. Elephant seal bulls often are unable to feed during the period when they are defending their territory during the breeding season. Young men engage in displays of bravado and risky social behaviour in the presence of peers and potential mates. Yet, there is no choice in such matters since mechanisms that enhance the perpetuation of genes influence behaviour devoid of conscious control. At the proximate level, in the *final common path* for behaviour (McFarland & Sibly, 1975), the signal for sexual behaviour carries more weight than feeding or drinking or exploring.

BEHAVIOUR IN A CONTEXT

Many of the studies that were examined in the preceding chapters have clearly indicated the importance of looking at the context in which motivated behaviour occurs. This principle was illustrated in Falk's (1972) research on adjunctive behaviour showing environmental control through the options available to animals placed on an intermittent feeding schedule. In Chapter 6 the similarities are discussed between adjunctive behaviours studied by psychologists and the displacement activitities observed and documented by ethologists. In both situations, the interruption of a consummatory behaviour in an intensely motivated animal induces the occurrence of another behaviour. This new behaviour immediately follows the interruption, and appears to be facilitated by environmental stimuli. For example, Zebra finches show different displacement activities depending upon the current stimulus situation. When it is threatened, it will feed if food is nearby, mount a female if one is present, or, if neither stimuli are available, it will preen or assume a sleeping posture (Morris, 1954).

The importance of context is also demonstrated in the work pioneered by Valenstein, Cox & Kakolewski (1968) on experimenter-induced consummatory responding by electrical stimulation discussed in Chapter 8. Their findings suggest that environmental control of behaviour is greater than most physiologically oriented researchers had originally assumed. Knowledge about the physiological state of the organism is not enough. It appears that environmental stimuli play an important role in determining the type of behaviour exhibited by an animal. The overlap in the neuroanatomical sites responsible for different forms of elicited behaviour is considerable (Valenstein, 1973). However, physiological plasticity may be restricted to behaviours subsumed under an 'appetitive' or 'aversive' reward system. Although there may be common sites underlying feeding, drinking, gnawing and stimulus seeking, the sites underlying reactions to aversive stimulation are quite distinctive.

INTERACTION OF EXTERNAL STIMULI, BEHAVIOUR AND PHYSIOLOGY

The material covered in the preceding chapters indicate another common theme – that different motivated behaviour patterns are the products of the interaction between the organism and its environment. In the analyses of individual motivational systems covered in each chapter, it is evident that there is reciprocal interaction between physiological processes and response patterns. I describe research on the courtship and parental activities to illustrate this point. Lehrman's (1964) work on courtship in ring doves revealed a sequence of events involving environmental stimulation, hormonal secretions and individual behaviour which influences hormonal secretions in another conspecific. Seasonal factors lead to bow-cooing in the male, the sight of which stimulates the female to become receptive as a result of endocrine changes. Courtship behaviour stimulates oestrogen release through hypothalamic pathways and provides the motivational basis for nest building in ring doves. The stimuli from male courtship activities cause the female to secrete oestrogen and progesterone. This results in her sexual receptivity. Lehrman found that oestrogen-injected doves would immediately start to build nests, and hypothesised that the nest-building activity and the sight of the nest stimulate the release of progesterone. The latter triggers ovulation in the females and motivates incubation behaviour. Injections of progesterone into isolated female and male doves once a day for seven days induce them to incubate immediately when presented with a nest of eggs.

Parental feeding of squabs is also influenced by behavioural and endocrine interactions of female and male doves. Lehrman (1955) found that injection of prolactin into ring doves causes the enlargement of the crop, which normally takes place before the young hatch. The crop growth prepares the adult birds to feed the squabs as a result of their capacity to secrete a milky fluid, but does not affect incubation appreciably. Lehrman hypothesised that the sight of the male sitting on the eggs stimulates secretion of prolactin in the female dove. The induction of parental feeding occurs as a result of sensory stimulation from the enlarged crop. Lehrman found that if afferent stimulation to the crop is cut off as a result of local anaesthetic applied to this region, the number of adults that feed squabs under such a condition is reduced. Control injections of the anaesthetic in other regions do not impair parental feeding.

Another variable that influences crop growth and related behaviour is stimulation provided by the squab. Hansen (1971) studied the effects of introducing newly hatched and eight-day-old squabs to experienced ring dove foster parents. These birds do not show specificity of parental care common among the precocial species discussed in Chapter 3. Squabs that were permitted to stay with the foster parents for eight successive days after hatching showed the greatest weight gain, and their foster parents also manifested crop growth. The results with the older squabs indicated that the foster parents were less responsive. Hansen suggested that continual change in crop size after hatching is a product of two factors. The first emanates from the physical appearance of the squab and other factors associated with attachment to the nest. The other concerns stimulus factors resulting from begging by and feeding of the squab.

Experience also influences hormonally mediated parental behaviour in ring doves. Lehrman & Wortis (1967) compared the reproductive and parental efficiency of ring doves that were breeding for the first time with those with previous breeding experience. The latter laid eggs sooner, incubated more efficiently and raised their young more successfully (indicated by rearing more squabs to maturity). When progesterone was injected into the inexperienced and experienced doves, there were differences in the frequency with which the hormone induced incubation behaviour. In general, 75 per cent of the experienced birds sat on the eggs whilst only 20 per cent of the inexperienced ones responded likewise. However, Lehrman (1971) did not conclude that all hormone-induced reproductive and parental behaviour patterns are learned. These complex behaviour patterns are also regulated by physiological mechanisms. Although these mechanisms may develop independently of traditional forms of experience, the latter is nevertheless relevant to the manifestation of these behaviour patterns.

The interaction of environmental stimuli and sexual receptivity in humans may be illustrated in the way that social interactions amongst groups of women could regulate their ovarian cycle (McClintock, 1998 a, b). Menstrual synchrony is a facet of a phenomenon that encompasses various forms of the timing of spontaneous ovulatory cycles in adult women. The mediating mechanism is believed to pheromones, airborne chemical signals, that are released by an individual into an environment and which affect the physiology and behaviour of other members of the same species. Stern & McClintock (1998) investigated compounds that regulate a specific neuroendocrine mechanism in other people without them being conscious of the odours. In general, they found that the timing of ovulation of the recipient can be manipulated by odours from women at either the early or late phase of their menstrual cycle.

CONTROL SYSTEMS

We have seen how feeding, drinking, exploring, escaping, avoiding or attacking serve to maintain optimal conditions for the individual organism. In all these cases, physiological states are involved even though the situations are very different and vary in complexity. In principle, each of these motivated activities can be analysed in terms of homeostatic mechanisms which operate to maintain the system in a state of relative balance. Environmental events and learning mechanisms interact with internal states in such a manner that the behavioural outcome is an expression of these factors. A feedback (and sometimes, feedforward) system is involved in all these motivated behaviours. Although feeding and drinking are not aroused simply by deficits, none the less, displacements in one direction do activate ingestion, and displacements in the other direction inhibit such activity. There are internal signals such as plasma glucose or cellular dehydration related to homeostatic control that set the levels of food or water intake. Anticipatory processes can also serve to regulate feedforward mechanisms. However, in the case of sexual behaviour, similar short-term homeostatic controls are absent. Nevertheless, the reward value of sexual behaviour has to be set so that it does occur sufficiently often to lead to successful reproductive behaviour (Rolls, 1999, p. 219). Several factors, such as hormonal signals in relation to the oestrous cycle, set the relative reward value of sexual behaviour. Rolls stated that when oestrogen level is high, the female rat's appetite for food is suppressed and sexual behaviour is rewarding.

It is possible to suggest analogies between regulatory systems influencing fear, exploration and attack. A sudden intrusion can instigate behaviour that attempts to restore the system to its prior state. The animal or person is hyperaroused by an aversive event and either removes itself from the source or removes the source. In the case of exploration, the individual responds on the basis of a stored representation of the environment, and is aroused when it perceives a mild disparity between incoming sensory information and its cognitive model of the environment (Toates, 1986). Much in the way that drinking can be re-aroused by hypertonic salt injection and feeding by insulin injection, aggression, fear and exploration can be aroused by appropriate stimuli, despite their recent performance.

INTERACTION BETWEEN MOTIVATED ACTIVITIES

Although each chapter in this book dealt with specific motivated acts, there is much interaction among them. Individual motivational systems dealing with, for example, reproduction, nutrition and defence, were analysed from the perspective of experiments that, by and large, deal with goal-seeking behaviour directed to only one goal. In the world outside of the controlled conditions in the laboratory, factors related to other goals are involved. On this point, we have observed the close relationship between feeding and drinking. The basis of their interaction comes from the fact that food intake creates a need for additional water intake through biochemical and osmotic processes. If protein is ingested (see discussion in Chapter 6), the urea formed requires water to flush it away, and the ingestion of dry food pulls water into the intestine osmotically, resulting in a systemic water deficit. Feeding and drinking are also closely associated in time, due partly to synchronised circadian rhythms in the two activities (see Oatley, 1971). Because eating creates a need for water, water-deprived rats reduce their intake of dry food in a quantitatively precise fashion (Collier & Knarr, 1966). The equivalent of about 1 ml of thirst reduces food intake by 0.33 g, whether thirst is induced by deprivation or by salt injection (Oatley & Toates, 1973). The depression of feeding arising from thirst depends upon the inhibitory processes within the nervous system.

In Chapter 7 the complex relationship between the motivational states of fear and exploration is discussed and experiments examined that indicate specific circumstances where mild fear may facilitate exploration even though, normally, fearful animals are cautious in returning to situations formerly associated with an aversive experience. There are similar complex interactions involved when a hungry animal is faced with a novel food source or a familiar food in a novel location. In this situation a motive for feeding will promote ingestion, whereas a motive of fear may promote the goal of keeping away from the novel source. Feeding and fear tendencies pull in opposite directions. The food-deprived animal shows food neophobia to avoid the possibility of eating something that may be tainted. Alternatively, the presence of a familiar and, presumably, 'safe' food in a novel location might prove to be a trap. With increased exposure to the novel food or location, animals habituate to the novelty, and show an increasing tendency to ingest rather than avoid the food. Children show a strong preference for familiar foods and are neophobic to most edibles that are outside of their range of experience (Birch, 1990), and exposure to different foods increases preferences for them.

Another interesting interaction of fear and exploration may be seen from the results of Blanchard, Kelley & Blanchard's study (1974) on defensive reactions and exploratory behaviour in rats. They found that when rats were given forced pre-exposure to a novel situation, there was a reduction in their subsequent latencies to leave a home cage and enter the novel situation. These researchers interpreted their results to indicate that novel situations produce a pattern of defensive reactions very similar to those elicited by aversive stimuli such as electric shock. Aversive stimuli produce either freezing or attempts at escape or avoidance, depending upon the features of the stimulus and the situation in which the aversive stimulus is presented. In the free-exploration task used in the experiment, the control rats rats showed a higher freezing rate in the exploratory alley and a greater tendency to remain in the home cage than those given forced exposure in the exploratory alley. Because of the differential exposure to the situation, the experimental rats found the stimuli in the exploratory alley less aversive than did the control rats. In general, these results indicate that novel situations produce a complex pattern of defensive reactions and that independent variables which produce changes in stimulus seeking may have their effect through the modification of one or more specific defensive reactions such as flight or freezing.

GOAL-ORIENTED BEHAVIOUR AND PHYSIOLOGICAL REDUCTIONISM

Motivated behaviour is characterised by goal-directed commerce with incentives (Toates, 1980; 1986). The notion that there is such a thing as truly goal-directed, intentional behaviour in animals or even in people has been questioned by McFarland (1989). It is true that behaviour cannot be controlled by foreknowledge of its ends, since an intended end often fails to come about. However, behaviour could be controlled by a psychological representation of some end. Animals and humans are quite capable of acquiring learned associations between a given behaviour and a certain outcome. There is a vast literature on Pavlovian conditioning that makes this point indisputable. The proximate causal mechanisms of motivated behaviours involve reactions to stimuli whose salience are of biological significance, thus eliciting withdrawal if unpleasant or approach if pleasant. Stimuli that have acquired negative or positive value as a result of the conditioning process act as incentives, and also elicit withdrawal or approach reactions, respectively.

The causal mechanisms of behaviour form an unbroken continuum of integrative levels. Psychologists and neurophysiologists differ on the level of the phenomenon being studied but not in other respects. The position taken throughout this book is that motivated behaviour is regarded as the product of an intimate interaction between the organism and its environment. In essence, physiological and behavioural events are treated as co-equal variables. 'Behaviour is not simply a projection of events at lower levels of the hierarchy, but behavioural interactions nevertheless arise from physiological integration' (Kennedy, 1992). The reciprocal interaction of physiological processes and response patterns was illustrated by my earlier analysis of the causes of reproductive and parental activities in the ring dove. The behavioural event involves the participation of the total organism, not merely special organs or tissues (Lazar, 1974). In addition, we would not find out about an animal's behaviour if the first thing we did was to record from its brain with no knowledge at all of what it could do (Dawkins, 1995). Unless the researcher understands what the animal can do through behavioural studies, poking around inside it may be fairly meaningless. Berridge's (1996) neurophysiological studies revealing vastly different systems mediating appetite for food (*wanting*) and the palatability/pleasure of it (*liking*) came about as a result of his success in developing behavioural measures that separated these components of food reward. As a result of this conceptual distinction, he uncovered separable underlying brain substrates that mediated these psychological reactions. This work is discussed in the following section.

AFFECT AND MOTIVATION

Humans want the rewards they like and like the rewards they want. Although this axiom is generally true, Berridge (1996) argued that *wanting* and *liking* do not always go together. He suggested that *liking* corresponds to the psychological concept of *palatability* whilst *wanting* corresponds more closely to *appetite* or *craving*. In some respects, the separate motivational mechanisms for generating *wanting* and *liking* corresponds to the *preparatory* and *consummatory* aspects of motivated behaviour, respectively. To distinguish between appetite and palatability implies the distinction between the disposition to eat and the sensory pleasure of actually eating. This is a distinction between motivation and affect which was difficult to test because traditional measures of palatability involved the quantity eaten or operant responding. They measured actions that are performed in order to obtain the next reward, whether the action is to press a bar or simply initiate the next bite of an external food. However, the development of Grill & Norgren's (1978) method of measuring affective reaction patterns arising from contact with a food, provided Berridge with a measure that directly reflects the liking or palatability of the test stimulus. This method was useful in demonstrating different taste reaction patterns among sodium-depleted and control rats ingesting a saline solution (Berridge *et al.*, 1984) as described in Chapter 5.

If wanting and liking are separable psychological processes with separate neural substrates, this must be demonstrated in experiments. Berridge & Robinson (1998) found that dopamine systems mediate the *incentive salience* of rewards, and modulate their motivational value in a manner separable from their hedonic properties. These researchers studied the consequences of dopamine depletion in the nucleus accumbens and neostriatum and found normal hedonic reaction patterns to sweet and bitter tastes. This result and others on the role of dopamine on reward suggest that dopamine systems are not necessary for the mediation of the hedonic pleasure of reward, but instead are more important to incentive saline attributions to the neural representations of reward-related stimuli. These factors constitute the mechanism underlying wanting. In contrast, the liking component of reward is dependent upon the integrity of opioid and benzodiazepine/GABA neurotransmitter systems as well as a specific region near the lateral hypothalamus but outside of it (Berridge, 1996).

The dissociation of mechanisms mediating wanting and liking has also been the outcome of Wise's (1994) review of the status of his recent experiments on neuroleptic drugs. Wise's (1982) initial work led him to suggest a role for dopamine in brain stimulation, psychomotor stimulant, food, water and opiate reinforcement from which he proposed the anhedonia hypothesis. He hypothesised that dopamine receptor antagonists (neuroleptic drugs) block the positive reinforcement and positive affect that is associated with rewarding events. Wise (1994) has revised his views and concluded that the anhedonia hypothesis wrongly linked subjective pleasure with objective reward. Part of the reason for this change of view arose from research indicating the dissociation between the rewarding and hedonic effects of cocaine and nicotine. This phenomenon was analysed from a similar perspective by Berridge & Robinson (1998) whose work is described above. Wise (1994) proposed that the hedonic and rewarding effects of appetite events must be separated, and that dopamine is mainly implicated in the mediation of psychomotor arousal or activation. Thus, independent research from Berridge's and Wise's laboratories suggest that wanting and liking are mediated by different physiological mechanisms.

There may be implications of Berridge's works to the understanding of rewards other than food. It is plausible that water deprivation evokes a wanting of liquid which would enable the animal to restore its water balance, and the taste of water evokes reactions corresponding to liking. Such a relationship was established in sodium-motivated behaviour. Sodium-depleted rats will work for access to this substance (Dickinson, 1986; Krieckhaus & Wolf, 1971; Weisinger, Woods & Skorupski, 1970) and respond positively to its taste in a taste reactivity test (Berridge *et al.*, 1984). The wanting and liking distinction may be useful in analysing why gerbils and hamsters ingest bitter-flavoured seeds or nuts as quickly as they do unflavoured edibles (Wong, 1994). Almost all other mammals inhibit their intake of bitter-flavoured foods. Presumably, there is a correspondence between intake and affect in other animals, but whether there is a dissociation between these processes in gerbils and hamsters is an issue which can be resolved with Berridge's analysis.

The availability of an opportunity to explore a novel situation may elicit wanting and liking of such activities. This is most likely if the novel stimulus is mildly discrepant with the features of the environment. However, neophobia is likely when the novel stimulus is greatly divergent from stimuli to which the animal had been exposed. What about sexual motivation? One may make the case that in humans, there are separable wanting and liking components of sexual behaviour. It is more difficult to argue this point in non-human animals that do not experience orgasm. Similarly, one could argue that there are both appetitive as well as affective aspects of parental caregiving activities. The mother is motivated to provide caregiving behaviour because of selection of mechanisms that facilitate it proximally, but also responds affectively to its presence or absence. This was elaborated in Chapter 3.

The preceding discussion certainly suggests that affect and appetite accompany the manifestation of motivated behaviour. Although the effects of these processes may be dissociated through refined behavioural techniques and there may be separable neural substrates underlying them, their expression is interactive. Affective reactions accompany the expression of all motivated actions. One exception may be the phenomenon of drug addiction which is an example of motivated behaviour where there is striving (goal directed activity) without the experience of pleasure or pain. This happens amongst long-term drug addicts who, as their addiction becomes established, experience a craving for the drug even though their experience of pleasure from it decreases (Berridge & Robinson, 1998). This craving is seen most dramatically from the use of stimulant drugs such as cocaine and amphetamines, where addiction is characterised by binges of continuous use, despite a tolerance of the euphoric effects that develops during the first few doses of a binge. Berridge & Robinson (1998) posit that the activation of the mesolimbic dopamine pathway by drugs mediates drug craving and addiction, but not drug induced pleasure. The latter arises from activation of the opioid and GABA neurotransmitter systems.

Motivated activities such as mating, parental care, feeding, drinking and exploring produce feedback that have positive hedonic consequences. Such activities are continued until the feedback produces negative hedonic reactions, and the activity is terminated. Behaviours motivated by negative hedonic stimuli such as pain, threat, danger, bad smells and the like threaten the integrity of the body and perhaps, individual survival. Hence, behaviours that remove the organism from the source serves as a critical part of its repertoire. Learning or associative processes may provide the mechanism for the anticipation of aversive events and promote adaptive avoidance reactions. Whether or not the causal mechanisms of all motivated activities involve separable wanting and liking components, positive and negative feedback do contribute to their manifestation.

Cabanac (1971) proposed that sensory pleasure tags stimuli that are useful for optimal physiological functioning and survival, and obversely, that sensory displeasure indicates useless or noxious stimuli. On the basis of McFarland & Sibly's (1975) notion of behaviour as a *final common path,* Cabanac (1996) suggested that a common currency must exist within the brain that allows trade-offs between motivations which result in ranking by order of priority. Such considerations and the results of a series of experiments designed to test this notion led Cabanac (1996) to conclude that pleasure is the common currency, and that it is a force that orientates behaviour toward stimuli associated with it. This process may be a part of all consummatory behaviours. Similarly, Rolls (1999, p. 275) suggested that 'reward and punishment signals provide a common currency for different sensory inputs, and can be seen as important in the selection of which actions are performed. Evolution ensures that the different reward and punishment signals are made potent to the extent that each will be chosen when appropriate'.

SUMMARY

Motivated behaviours are goal-oriented and may be bifurcated by two processes, the wanting and the liking of the reward associated with goal-attainment. The functional significance of motivated activities is viewed in terms of fitness maximisation. Behaviours occur in an environmental context producing interactions of external stimuli, behaviour and physiology. Control systems link environmental events and internal states with the linkage facilitated by learning. Such systems adaptively mediate between environmental events and internal states by generating multiple, mutually interacting motivated activities, all which have as common currency – pleasure. Sensory pleasure tags stimuli that are useful for optimal physiological functioning and survival, whilst sensory displeasure indicates noxious stimuli.

References

Abdelaal, A. E., Mercer, P. F. & Mogenson, G. J. (1976). Plasma angiotensin II levels and water intake following beta-adrenergic stimulation, hypovolemia, cellular dehydration and water deprivation. *Pharmacology, Biochemistry and Behavior,* **4**, 317–21.

Aberle, D. F. (1961). Navaho. In *Matrilineal Kinship*, ed. D. M. Schneider & K. Gough, pp. 96–201. Berkeley: University of California Press.

Abraham, S. F., Denton, D. A. & Weisinger, R. S. (1976). The specificity of the dipsogenic effect of angiotensin. *Pharmacology, Biochemistry and Behavior,* **4**, 363–8.

Adam, W. R. (1973). Novel diet preferences in potassium-deficient rats. *Journal of Comparative and Physiological Psychology,* **84**, 286–8.

Adam, W. R. & Dawborn, J. K. (1972). Effect of potassium depletion on mineral appetite in the rat. *Journal of Comparative and Physiological Psychology,* **78**, 51–8.

Adams, D. & Flynn, J. P. (1966). Transfer of an escape response from tail shock to brain-stimulated attack behavior. *Journal of the Experimental Analysis of Behavior,* **9**, 401–8.

Ader, R. & Friedman, S. B. (1965). Differential early experience and susceptibility to transplanted tumor in the rat. *Journal of Comparative and Physiological Psychology,* **59**, 361–4.

Ader, R. & Grota, L. J. (1969). Effects of early experience on adrenocortical reactivity. *Physiology and Behavior,* **4**, 303–5.

Adolph, E. F. (1939). Measurement of water drinking in dogs. *American Journal of Physiology,* **125**, 75–86.

Ahern, G. L., Landin, M. L. & Wolf, G. (1978). Escape from deficits in sodium intake after thalamic lesions as a function of preoperative experience. *Journal of Comparative and Physiological Psychology,* **92**, 544–54.

Ainsworth, M. D., Blehar, M. C., Waters, E. & Walls, S. (1978). *Patterns of attachment: a psychological study of the strange situation.* Hillsdale, NJ: Lawrence Erlbaum Associates.

Alcock, J. (1998). *Animal behavior: an evolutionary approach.* Sunderland, MA: Sinauer Associates.

Alexander, G. & Shillito, E. (1977). The importance of odour appearance and voice in maternal recogntion in Merino sheep. *Applied Animal Ethology,* **7**, 127–35.

Altman, J. & Das, G. D. (1965). Autoradiographic and histological evidence of postnatal hippocampal neurogenesis in rats. *Journal of Comparative Neurology,* **124**, 319–36.

Altman, J., Das, G. D. & Anderson, W. J. (1968). Effects of infantile handling on morphological development of the rat brain: an exploratory study. *Developmental Psychobiology*, **1**, 10–29.

Amsel, A. (1962). Frustrative nonreward in partial reinforcement and discrimination learning: some recent history and a theoretical extension. *Psychological Review*, **69**, 306–28.

Anand, B. K. & Brobeck, J. R. (1951). Hypothalamic control of food intake. *Yale Journal of Biology and Medicine*, **24**, 123–40.

Anderson, P. (1983). The reproductive role of the human breast. *Current Anthropology*, **24**, 25–45.

Andersson, M. (1982). Female choice selects for extreme tail length in a widowbird. *Nature*, **299**, 818–20.

Andreas, P., Kuester, J. & Arnemann, J. (1992). DNA fingerprinting reveals that infant care by male Barbary macaques (*Macaca sylvanus*) is not parental investment. *Folio Primatology*, **58**, 93–8.

Aravich, P. F., Doerries, L. E. & Rieg, T. S. (1994). Exercise-induced weight loss in the rat and anorexia nervosa. *Appetite*, **23**, 196.

Arnold, A. P. (1996). Genetically triggered sexual differentiation of brain and behavior. *Hormones and Behavior*, **30**, 495–505.

Axelrod, R. & Hamilton, W. D. (1981). The evolution of cooperation. *Science*, **211**, 1390–6.

Azrin, N. H., Hutchison, R. R. & McLaughlin, R. (1965). The opportunity for aggression as an operant reinforcer during aversive stimulation. *Journal of the Experimental Analysis of Behavior*, **7**, 223–8.

Baird, A. A., Gruber, S. A., Fein, D. A., Maas, L. C., Steingard, R. J., Renshaw, P. F., Cohen, B. M. & Yurgelun-Todd, D. A. (1999). Functional magnetic resonance imaging of facial affect recognition in children and adolescents. *Journal of the American Academy of Child and Adolescent Psychiatry*, **38**, 195–9.

Baker, B. J. & Booth, D. A. (1989). Genuinely olfactory preferences conditioned by protein repletion. *Appetite*, **13**, 223–7.

Balagura, S. & Devenport, L. D. (1970). Feeding patterns of normal and ventromedial hypothalamic lesioned male and female rats. *Journal of Comparative and Physiological Psychology*, **71**, 357–64.

Barnes, G. W. & Kish, G. B. (1958). On some properties of visual reinforcement. *American Psychologist*, **13**, 417.

Barnett, S. A. (1956). Behaviour components in the feeding of wild and laboratory rats. *Behaviour*, **9**, 24–43.

Barnett, S. A. (1958). Experiments in 'neophobia' in wild and laboratory rats. *British Journal of Psychology*, **49**, 195–201.

Barnett, S. A. & Cowan, P. E. (1976). Activity, exploration, curiosity and fear: an ethological study. *Interdisciplinary Science Reviews*, **1**, 43–62.

Bartoshuk, L. (1968). Water taste in man. *Perception and Psychophysics*, **3**, 69–72.

Bartoshuk, L. M. (1989). The functions of taste and olfaction. *Annals of the New York Academy of Sciences*, **575**, 353–61.

Baskin, D. G., Wilcox, B. J., Figlewicz, D. P. & Dorsa, D. M. (1988). Insulin and insulin-like growth factors in the CNS. *Trends in Neurosciences*, **11**, 107–11.

Batten, M. (1992). *Sexual strategies: how females choose their mates.* New York: G.P. Putnam's Sons.

Baura, G. D., Foster, D. M., Porte, D., Jr., Kahn, S. E., Bergman, R. N., Cobelli, C. & Schwartz, M. W. (1993). Saturable transport of insulin from plasma into the central nervous system of dogs in vivo. A mechanism for regulated insulin delivery to the brain. *Journal of Clinical Investigations*, **92**, 1824–30.

Beach, F. A. & Jaynes, J. (1956). Studies of maternal retrieving in rats. I. Recognition of young. *Journal of Mammal*, **37**, 177–88.

Beach, F. A., Noble, R. G. & Orndoff, R. K. (1969). Effects of perinatal androgen treatment on responses of male rats to gonadal hormones in adulthood. *Journal of Comparative and Physiological Psychology*, **68**, 490–7.

Beatty, W. W. (1979). Gonadal hormones and sex differences in nonreproductive behaviors in rodents: organizational and activational influences. *Hormones and Behavior*, **12**, 112–63.

Beatty, W. W. & Beatty, P. A. (1970). Hormonal determinants of sex differences, avoidance behavior and reactivity to electric shock in the rat. *Journal of Comparative and Physiological Psychology*, **73**, 446–55.

Beauchamp, G. K., Bertino, M., Burke, D. & Engelman, K. (1990). Experimental sodium depletion of salt taste in normal human volunteers. *The American Journal of Clinical Nutrition*, **51**, 881–9.

Beck, B., Burlet, A., Nicolas, J. P. & Burlet, C. (1990). Hypothalamic neuropeptide Y (NPY) in obese Zucker rats: implications in feeding and sexual behaviors. *Physiology and Behavior*, **47**, 449–53.

Beck, M., Hitchcock, C. L. & Galef, B. G. (1988). Diet sampling by wild Norway rats offered several unfamiliar foods. *Animal Learning & Behavior*, **16**, 224–30.

Beck, R. C. (1978). *Motivation: theories and principles*. Englewood Cliffs: Prentice-Hall.

Beck, R. C. (1990). *Motivation: theories and principles*. 2nd edn. Englewood Cliffs: Prentice-Hall.

Bellows, R. T. (1939). Time factors in water drinking in dogs. *American Journal of Physiology*, **125**, 87–97.

Bemis, K. M. (1978). Current approaches to the etiology and treatment of anorexia nervosa. *Psychological Bulletin*, **85**, 593–617.

Bercovitch, F. B. (1988). Coalitions, cooperation and reproductive tactics among adult male baboons. *Animal Behaviour*, **36**, 1198–209.

Berenbaum, S. A. (1999). Effects of early androgens on sex-typed activities and interests in adolescents with congenital adrenal hyperplasia. *Hormones and Behavior*, **35**, 102–10.

Berenbaum, S. A. & Snyder, E. (1995). Early hormonal influences on childhood sex-typed activity and playmate preferences. *Developmental Psychology*, **31**, 31–42.

Berkowitz, L. (1983). Aversively stimulated aggression. Some parallels and differences in research with animals and humans. *American Psychologist*, **38**, 1135–44.

Berkowitz, L. (1989). Laboratory experiments in the study of aggression. In *Human aggression: naturalistic approaches*, ed. J. Archer & K. Browne, pp. 42–61. London: Routledge.

Berlyne, D. E. (1960). *Conflict, arousal, and curiosity*. New York: McGraw-Hill.

Berlyne, D. E. (1963). Motivational problems raised by exploratory and epistemic behavior. In *Psychology: a study of a science*, ed. S. Koch, pp. 284–364. New York: McGraw-Hill.

Berlyne, D. E. (1967). Arousal and reinforcement. In *Nebraska symposium on motivation*, ed. D. Levine, pp. 1–110. Lincoln: University of Nebraska Press.

Bernard, C. (1879). *Leçons de physiologie opératoire*. Paris: Baillière.

Bernardis, L. L. & Bellinger, L. L. (1996). The lateral hypothalamic area revisited: ingestive behavior. *Neuroscience and Biobehavioral Reviews*, **20**, 189–287.

Berrebi, A. S., Fitch, R. H., Ralphe, D. L., Denenberg, J. O., Friedrich, V. L., Jr. & Denenberg, V. H. (1988). Corpus callosum: region-specific effects of sex, early experience and age. *Brain Research*, **438**, 216–24.

Berridge, K. C. (1996). Food reward: brain substrates of wanting and liking. *Neuroscience and Biobehavioral Reviews*, **20**, 1–25.

Berridge, K. C., Flynn, F. W., Schulkin, J. & Grill, H. J. (1984). Sodium depletion enhances salt palatability in rats. *Behavioral Neuroscience*, **98**, 652–60.

Berridge, K. C. & Robinson, T. E. (1998). What is the role of dopamine in reward: hedonic impact, reward learning, or incentive salience? *Brain Research Reviews*, **28**, 309–69.

Berridge, K. C., Venier, I. L. & Robinson, T. E. (1989). Taste reactivity analysis of 6-hydroxydopamine-induced aphagia: implications for arousal and anhedonia hypotheses of dopamine function. *Behavioral Neuroscience*, **103**, 36–45.

Bindra, D. (1959). *Motivation, a systematic reinterpretation*. New York: Ronald Press.

Bindra, D. (1969). A unified interpretation of emotion and motivation. *Annals of the New York Academy of Sciences*, **159**, 1071–83.

Bindra, D. (1974). A motivational view of learning, performance, and behavior modification. *Psychological Review*, **81**, 199–213.

Bindra, D. (1978). How adaptive behavior is produced: a perceptual-motivational alternative to response reinforcement. *Behavioral and Brain Sciences*, **2**, 41–91.

Binnie-Dawson, J. L. & Cheung, Y. M. (1982). The effects of different types of neonatal feminization and environmental stimulation on changes in sex-associated activity/spatial learning skills. *Biological Psychology*, **15**, 109–40.

Birch, L. (1980). Effects of peer models' food choices and eating behaviors on preschoolers' food preferences. *Child Development*, **51**, 489–96.

Birch, L. (1990). The control of food intake by young children. In *Taste, experience and feeding*, ed. E. D. Capaldi & T. L. Powley, pp. 115–35. Washington, DC: American Psychological Association.

Birch, L., Zimmerman, S. I. & Hind, H. (1980). The influence of social-affective context on the formation of children's food preferences. *Child Development*, **51**, 856–61.

Birch, L. L., Birch, D., Marlin, D. W. & Kramer, L. (1982). Effects of instrumental consumption on children's food preference. *Appetite*, **3**, 125–34.

Birch, L. L. & Marlin, D. W. (1982). I don't like it; I've never tried it; effects of exposure on two-year-old children's food preferences. *Appetite*, **3**, 353–60.

Birch, L. L., McPhee, L., Shoba, B. C., Steinberg, L. & Krehbiel, R. (1987). 'Clean up your plate': effects of child feeding practices on the conditioning of meal size. *Learning & Motivation*, **18**, 301–17.

Birke, L. I. & Sadler, D. (1985). Maternal behavior in rats and the effects of neonatal progestins given to the pups. *Developmental Psychobiology*, **18**, 467–75.

Birkhead, T. R. (1996). Sperm competition: evolution and mechanisms. *Current Topics in Developmental Biology*, **33**, 103–58.

Birkhead, T. R., Burke, T., Zann, R., Hunter, F. M. & Krupa, A. P. (1990). Extra-pair paternity and intraspecific brood parasitism in wild zebra finches *Taeniopygia guttata* revealed by DNA fingerprinting. *Behavioural Ecology and Sociobiology*, **27**, 315–24.

Birkhead, T. R., Clarkson, K. & Zann, R. (1988). Extra-pair courtship, copulation, and mate guarding in wild zebra finches, *Taeniopygia guttata*. *Animal Behaviour*, **35**, 1853–5.

Blackburn, J. R., Phillips, A. G. & Fibiger, H. C. (1987). Dopamine and preparatory behavior: I. Effects of pimozide. *Behavioral Neuroscience*, **101**, 352–60.

Blackburn, J. R., Phillips, A. G., Jakubovic, A. & Fibiger, H. C. (1989). Dopamine and preparatory behavior: II. A neurochemical analysis. *Behavioral Neuroscience*, **103**, 15–23.

Blanchard, R. J., Kelley, M. J. & Blanchard, D. C. (1974). Defensive reactions and exploratory behavior in rats. *Journal of Comparative and Physiological Psychology*, **87**, 1129–33.

Blass, E. M. (1974). Evidence for basal forebrain thirst osmoreceptors in rat. *Brain Research*, **82**, 69–76.

Blass, E. M. & Epstein, A. N. (1971). A lateral preoptic osmosensitive zone for thirst in the rat. *Journal of Comparative and Physiological Psychology*, **76**, 378–94.

Bleichfeld, B. & Moely, B. E. (1984). Psychophysiological responses to an infant cry: comparison of groups of women in different phases of the maternal cycle. *Developmental Psychology*, **20**, 1082–91.

Blundell, J. E. & Hill, A. J. (1992). Dexfenfluramine and appetite in humans. *International Journal of Obesity and Related Metabolic Disorders*, **16** (Suppl 3), S51–9.

Blundell, J. E. & Latham, C. J. (1979). Serotonergic influences on food intake: effect of 5-hydroxytryptophan on parameters of feeding behaviour in deprived and free-feeding rats. *Pharmacology, Biochemistry and Behavior*, **11**, 431–7.

Blundell, J. E. & Lawton, C. L. (1995). Serotonin and dietary fat intake: effects of dexfenfluramine. *Metabolism*, **44**, 33–7.

Boakes, R. A. & Dwyer, D. M. (1997). Weight loss in rats produced by running: effects of prior experience and individual housing. *Quarterly Journal of Experimental Psychology*, **50B**, 129–48.

Bolles, R. C. (1967). *Theory of motivation*. New York: Harper & Row.

Bolles, R. C. (1970). Species-specific defense reactions. *Psychological Review*, **77**, 32–48.

Bolles, R. C. (1979). Toy rats and real rats: nonhomeostatic plasticity in drinking. *Behavioral and Brain Sciences*, **2**, 103.

Bolles, R. C. & de Lorge, J. (1962). The rat's adjustment to a-diurnal feeding cycles. *Journal of Comparative and Physiological Psychology*, **83**, 510–14.

Bookstein, F. L. (1993). Converting cultural success into mating failure by aging. *Behavioral and Brain Sciences*, **12**, 285–6.

Booth, D. A. (1994). *The psychology of nutrition*. Bristol, PA: Taylor and Francis.

Booth, D. A. & Simson, P. C. (1971). Food preferences acquired by association with variations in amino acid nutrition. *Quarterly Journal of Experimental Psychology*, **23**, 135–45.

Borsini, F. & Rolls, E. T. (1984). Role of noradrenaline and serotonin in the basolateral region of the amygdala in food preferences and learned taste aversions in the rat. *Physiology and Behavior*, **33**, 37–43.

Bowlby, J. (1969). *Attachment and loss*. London: Hogarth.

Bowlby, J. (1989). The role of attachment in personality development and psychopathology. In *The course of life*, ed. S. I. Greenspan & G. H. Pollock, pp. 229–70. Madison: International Universities Press.

Bowlby, J. (1991). Ethological light on psychoanalytical problems. In *The development and integration of behaviour: essays in honour of Robert Hinde*, ed. P. Bateson, pp. 301–13. Cambridge: Cambridge University Press.

Box, H. O. (1977). Quantitative data on the carrying of young captive monkeys (Calithrix jacchus) by other members of their family groups. *Primates*, **18**, 475–84.

Braveman, N. S. & Jarvis, P. S. (1978). Independence of neophobia and taste aversion. *Animal Learning and Behavior*, **6**, 406–12.

Brett, L. P. & Levine, S. (1979). Schedule-induced polydipsia suppresses pituitary-adrenal activity in rats. *Journal of Comparative and Physiological Psychology*, **93**, 946–56.

Bridges, R. S. (1984). A quantitative analysis of the roles of dosage, sequence, and duration of estradiol and progesterone exposure in the regulation of maternal behavior in the rat. *Endocrinology*, **114**, 930–40.

Bridges, R. S. & Mann, P. E. (1994). Prolactin-brain interactions in the induction of material behavior in rats. *Psychoneuroendocrinology*, **19**, 611–22.

Bridges, R. S., Numan, M., Ronsheim, P. M., Mann, P. E. & Lupini, C. E. (1990). Central prolactin infusions stimulate maternal behavior in steroid- treated, nulliparous female rats. *Proceedings of the National Academy of Sciences USA*, **87**,

8003–7.
Bridges, R. S., Robertson, M. C., Shiu, R. P., Sturgis, J. D., Henriquez, B. M. & Mann, P. E. (1997). Central lactogenic regulation of maternal behavior in rats: steroid dependence, hormone specificity, and behavioral potencies of rat prolactin and rat placental lactogen I. *Endocrinology,* **138**, 756–63.
Bronstein, P. M., Neiman, H., Wolkoff, F. D. & Levine, M. J. (1974). The development of habituation in the rat. *Animal Learning and Behavior,* **2**, 92–6.
Brown, J. S. (1961). *The motivation of behavior.* New York: McGraw-Hill.
Brown, R. T. & Wagner, A. R. (1964). Resistance to punishment and extinction following training with shock or nonreinforcement. *Journal of Experimental Psychology,* **68**, 503–7.
Brownell, K. D. (1982). Obesity: understanding and treating a serious, prevalent, and refractory disorder. *Journal of Consulting and Clinical Psychology,* **50**, 820–40.
Brownell, K. D., Greenwood, M. R., Stellar, E. & Shrager, E. E. (1986). The effects of repeated cycles of weight loss and regain in rats. *Physiology & Behavior,* **38**, 459–64.
Brownell, K. D. & Stunkard, A. J. (1978). Behavior therapy and behavior change: uncertainties in programs for weight control. *Behaviour Research and Therapy,* **16**, 301.
Bruch, H. (1973). *Eating disorders: obesity, anorexia nervosa, and the person within.* New York: Basic Books.
Brunjes, P. C. (1988). Precocity and plasticity: odor deprivation and brain development in the precocial mouse *Acomys cahirinus. Neuroscience,* **24**, 579–82.
Burnstein, E., Crandall, C. & Kitayama, S. (1994). Some neo-Darwinian decision rules for altruism: weighing cues for inclusive fitness as a function of the biological importance of the decision. *Journal of Personality and Social Psychology,* **67**, 773–89.
Burton, M. J., Rolls, E. T. & Mora, F. (1976). Effects of hunger on the responses of neurons in the lateral hypothalamus to the sight and taste of food. *Experimental Neurology,* **51**, 668–77.
Buss, D. M. (1989). Sex differences in human mate preferences: evolutionary hypothesis tested in 37 cultures. *Behavioral and Brain Sciences,* **12**, 1–49.
Buss, D. M. (1999). *Evolutionary psychology: the new science of the mind.* Boston: Allyn and Bacon.
Bunnk, B. P., Angleitner, A., Oubaid, V. & Buss, D. M. (1996). Sex differences in jealousy in evolutionary and cultural perspective. *Psychological Science,* **7** 359–63.
Cabanac, M. (1971). Physiological role of pleasure. *Science,* **173**, 1103–7.
Cabanac, M. (1996). On the origin of consciousness, a postulate and its corollary. *Neuroscience and Biobehavioral Reviews,* **20**, 33–40.
Cabanac, M. & Duclaux, R. (1970). Obesity: absence of satiety aversion to sucrose. *Science,* **168**, 496–7.
Caggiula, A. R. (1970). Analysis of the copulation-reward properties of posterior hypothalamic stimulation in male rats. *Journal of Comparative and Physiological Psychology,* **70**, 399–412.
Callard, G. V., Petro, Z. & Ryan, K. J. (1978). Phylogenetic distribution of aromatase and other androgen-converting enzymes in the central nervous system. *Endocrinology,* **103**, 2283–90.
Campbell, D. T. & Gatewood, J. B. (1994). Ambivalently held group-optimizing predispositions. *Behavioral and Brain Sciences,* **17**, 614.
Cannon, W. B. (1929). *A laboratory course in physiology.* Cambridge, MA: Harvard University Press.

Cantwell, D. P., Sturzenberger, S., Burroughs, J., Salkin, B. & Green, J. K. (1977). Anorexia nervosa. An affective disorder? *Archives of General Psychiatry*, **34**, 1087–93.

Capretta, P. J., Petersik, J. T. & Stewart, D. J. (1975). Acceptance of novel flavours is increased after early experience of diverse tastes. *Nature*, **254**, 689–91.

Carek, D. J. & Capelli, A. J. (1981). Mothers' reactions to their newborn infants. *Journal of the American Academy of Child Psychiatry*, **20**, 16–31.

Carlsson, S. G., Fagerberg, H., Horneman, G., Hwang, C. P., Larsson, K., Rodholm, M., Schaller, J., Danielsson, B. & Gundewall, C. (1978). Effects of amount of contact between mother and child on the mother's nursing behavior. *Developmental Psychobiology*, **11**, 143–50.

Caro, J. F., Sinha, M. K., Kolaczynski, J. W., Zhang, P. L. & Considine, R. V. (1996). Leptin: the tale of an obesity gene. *Diabetes*, **45**, 1455–62.

Carr, W. J. (1952). The effect of adrenalectomy upon NaCl taste threshold in the rat. *Journal of Comparative and Physiological Psychology*, **45**, 377–80.

Carroll, M. E., Dinc, H. I., Levy, C. J. & Smith, J. C. (1975). Demonstrations of neophobia and enhanced neophobia in the albino rat. *Journal of Comparative and Physiological Psychology*, **89**, 457–67.

Chitty, D. (1954). *Control of rats and mice.* London: Clarendon Press/Oxford University Press.

Clark, J. M., Clark, A. J., Bartle, A. & Winn, P. (1991). The regulation of feeding and drinking in rats with lesions of the lateral hypothalamus made by N-methyl-D-aspartate. *Neuroscience*, **45**, 631–40.

Clark, J. T., Kalra, P. S., Crowley, W. R. & Kalra, S. P. (1984). Neuropeptide Y and human pancreatic polypeptide stimulate feeding behavior in rats. *Endocrinology*, **115**, 427–9.

Clayton, D. H. (1990). Mate choice in experimentally parasitized rock doves: lousy males lose. *American Zoologist*, **30**, 251–62.

Clepet, C., Schafer, A. J., Sinclair, A. H., Palmer, M. S., Lovell-Badge, R. & Goodfellow, P. N. (1993). The human SRY transcript. *Human Molecular Genetics*, **2**, 2007–12.

Cleveland, J. & Snowdon, C. T. (1984). Social development during the first twenty weeks in the cotton-top tamarins (*Saguinus oedipus*). *Animal Behaviour*, **32**, 432–44.

Coburn, P. C. & Stricker, E. M. (1978). Osmoregulatory thirst in rats after lateral preoptic lesions. *Journal of Comparative and Physiological Psychology*, **92**, 350–61.

Coe, C. L., Mendoza, S. P., Smotherman, W. P. & Levine, S. (1978). Mother–infant attachment in the squirrel monkey: adrenal response to separation. *Behavioral Biology*, **22**, 256–63.

Colgan, P. W. (1989). *Animal motivation.* London: Chapman and Hall.

Collias, N. E. (1956). The analysis of socialization in sheep and goats. *Ecology*, **37**, 228–39.

Collier, G. & Knarr, F. (1966). Defense of water balance in the rats. *Journal of Comparative and Physiological Psychology*, **61**, 5–10.

Contreras, R. J. (1977). Changes in gustatory nerve discharges with sodium deficiency: a single unit analysis. *Brain Research*, **121**, 373–8.

Cooper, S. J., Dourish, C. T. & Clifton, P. G. (1992). CCK antagonists and CCK-monoamine interactions in the control of satiety. *American Journal of Clinical Nutrition*, **55**, 291S-5S.

Corbett, S. W. & Keesey, R. E. (1982). Energy balance of rats with lateral hypothalamic lesions. *American Journal of Physiology*, **242**, E273–9.

Corbit, J. D. & Stellar, E. (1964). Palatability, food intake and obesity in normal and hyperphagic rats. *Journal of Comparative and Physiological Psychology*, **58**, 63–7.

Corey, D. T. (1978). The determinants of exploration and neophobia. *Neuroscience and Biobehavioral Reviews*, **2**, 235–53.

Cosmides, L. & Tooby, J. (1987). From evolution to behavior: evolutionary psychology as the missing link. In *The latest on the best: essays on evolution and optimality*, ed. J. Dupre, pp. 276–306. Cambridge, MA: MIT Press.

Cosmides, L. & Tooby, J. (1992). Cognitive adaptations for social exchange. In *The adapted mind: evolutionary psychology and the generation of culture*, ed. J. H. Barkow, L. Cosmides & J. Tooby, pp. 163–228. New York: Oxford University Press.

Cosmides, L. & Tooby, J. (1995). From evolution to adaptations to behavior: toward an integrated evolutionary psychology. In *Biological perspectives on motivated activities*, ed. R. Wong, pp. 11–74. Norwood, NJ: Ablex Publishing.

Cox, J. E. & Powley, T. L. (1981). Prior vagotomy blocks VMH obesity in pair-fed rats. *American Journal of Physiology*, **240**, E573–83.

Crawford, C. B. & Anderson, J. L. (1989). Sociobiology: an environmentalist discipline? *American Psychologist*, **44**, 1449–59.

Crawley, M. J. (1983). *Herbivory, the dynamics of animal–plant interactions*. Berkeley: University of California Press.

Crnic, L., Reite, M. & Shucard, D. (1982). Animal models of human behavior. In *The development of attachment and affiliation*, ed. N. Emde & R. J. Harmon, pp. 31–42. New York: Plenum Press.

Cromwell, H. C. & Berridge, K. C. (1993). Where does damage lead to enhanced food aversion: the ventral pallidum/substantia innominata or lateral hypothalamus? *Brain Research*, **624**, 1–10.

Dalland, T. (1970). Response and stimulus perseveration in rats with septal and dorsal hippocampal lesions. *Journal of Comparative and Physiological Psychology*, **71**, 114–18.

Daly, M. & Wilson, M. (1983). *Sex, evolution, and behavior*. Boston: PWS Publishers.

Daly, M. & Wilson, M. (1988). *Homicide*. Hawthorne, NY: Aldine de Gruyter.

Daly, M. & Wilson, M. (1994). Evolutionary psychology of male violence. In *Male violence*, ed. J. Archer, pp. 253–88. London: Routledge.

Darwin, C. (1862). On the two forms, or dimorphic conditions in the species of Primula, and on their remarkable sexual relations. *Journal of the Proceedings of the Linnean Society (Botany)*, **6**, 45–63.

Darwin, C. (1871). *The descent of man, and selection in relation to sex*. New York: D. Appleton and Company.

Darwin, C. (1872). *The expression of the emotions in man and animals*. London: J. Murray.

Davis, C. M. (1928). Self-selection of diet by newly weaned infants. *American Journal of Diseases of Children*, **36**, 651–79.

Davis, C. M. (1939). Results of the self-selection of diets by young children. *Canadian Medical Association Journal*, **Sept**, 257–61.

Dawkins, M. S. (1986). *Unravelling animal behaviour*. Harlow: Longman.

Dawkins, M. S. (1995). *Unravelling animal behaviour*, 2nd edn. New York: Wiley.

Dawkins, R. (1976). *The selfish gene*. New York: Oxford University Press.

Dawkins, R. (1982). *The extended phenotype: the gene as the unit of selection*. San Francisco: Freeman.

Dawson, J. L. (1972). Effects of sex hormones on cognitive style in rats and men. *Behavior Genetics*, **2**, 21–42.

Dawson, J. L., Cheung, Y. M. & Lau, R. T. (1975). Developmental effects of neonatal sex hormones on spatial and activity skills in the white rat. *Biological Psychology*, **3**, 213–29.

de Toledo, L. & Black, A. H. (1966). Heart rate: changes during conditioned suppression in rats. *Science*, **152**, 1404–6.

deCatanzaro, D. (1999). *Motivation and emotion: evolutionary, physiological, developmental, and social perspectives.* Upper Saddle River, NJ: Prentice Hall.

Dember, W. N. & Earl, R. W. (1957). Analysis of exploration, manipulation and curiosity behavior. *Psychological Review*, **64**, 91–7.

Dember, W. N. & Richman, C. L. (1989). *Spontaneous alternation behavior.* New York: Springer-Verlag.

DeNelsky, G. Y. & Denenberg, V. H. (1967a). Infantile stimulation and adult exploratory behaviour in the rat: effects of handling upon visual variation-seeking. *Animal Behaviour*, **15**, 568–73.

DeNelsky, G. Y. & Denenberg, V. H. (1967b). Infantile stimulation and adult exploratory behavior: effects of handling upon tactual variation seeking. *Journal of Comparative and Physiological Psychology*, **63**, 309–12.

Denenberg, V. H. (1969). Open-field behavior in the rat: what does it mean? *Annals of the New York Academy of Sciences*, **159**, 852–9.

Denenberg, V. H. & Haltmeyer, G. C. (1967). Test of the monotonicity hypothesis concerning infantile stimulation and emotional reactivity. *Journal of Comparative and Physiological Psychology*, **63**, 394–6.

Denti, A. & Epstein, A. (1972). Sex differences in the acquisition of two kinds of avoidance behavior in rats. *Physiology and Behavior*, **8**, 611–15.

Denton, D. A. (1982). *The hunger for salt: an anthropological, physiological, and medical analysis.* New York: Springer-Verlag.

Deutsch, J. A. (1960). *The structural basis of behavior.* Chicago: University of Chicago Press.

Deutsch, J. A. (1985). The role of the stomach in eating. *American Journal of Clinical Nutrition*, **42**, 1040–3.

Deutsch, J. A. & Ahn, J. S. (1986). The splanchnic nerve and food intake regulation. *Behavioral and Neural Biology*, **45**, 43–7.

Deutsch, J. A., Moore, B. O. & Heinrichs, S. C. (1989). Unlearned specific appetite for protein. *Physiology and Behavior*, **46**, 619–24.

Diamond, M. C. (1988). *Enriching heredity: the impact of the environment on the anatomy of the brain.* New York: Free Press.

Dickinson, A. (1986). Re-examination of the role of the instrumental contingency in the sodium-appetite irrelevant incentive effect. *Quarterly Journal of Experimental Psychology*, **38B**, 161–72.

Doerries, L. E., Avarich, P. F., Metcalf, A., Wall, J. D. & Lauterio, T. J. (1989). β-endorphin and activity-based anorexia in the rat: influence of simultaneously initiated dieting and exercise on weight loss and -endorphin. *Annals of the New York Academy of Sciences*, **575**, 609–10.

Doty, R. L., Snyder, P. J., Huggins, G. R. & Lowry, L. D. (1981). Endocrine, cardiovascular, and psychological correlates of olfactory sensitivity changes during the human menstrual cycle. *Journal of Comparative and Physiological Psychology*, **95**, 45–60.

Douglas, R. J. (1967). The hippocampus and behavior. *Psychological Bulletin*, **67**, 416–22.

Douglas, R. J. (1989). Spontaneous alternation behavior and the brain. In *Spontaneous alternation behavior*, ed. W. N. Dember & C. L. Richman, pp. 73–108. New York: Springer-Verlag.

Draper, P. (1976). Social and economic constraints on child life among the !Kung. In *Kalahari hunter–gatherers*, ed. R. B. Lee & I. DeVore, pp. 218–45. Cambridge, MA: Harvard University Press.

Duggan, J. P. & Booth, D. A. (1986). Obesity, overeating, and rapid gastric emptying in rats with ventromedial hypothalamic lesions. *Science*, **231**, 609–11.

Dunnett, S. B., Lane, D. M. & Winn, P. (1985). Ibotenic acid lesions of the lateral hypothalamus: comparison with 6-hydroxydopamine-induced sensorimotor deficits. *Neuroscience,* **14**, 509–18.

Dworkin, B. R. (1993). *Learning and physiological regulation.* Chicago: University of Chicago Press.

Dwyer, D. M. & Boakes, R. A. (1997). Activity-based anorexia in rats as failure to adapt to a feeding schedule. *Behavioral Neuroscience,* **111**, 195–205.

Eberhard, W. G. (1996). *Female control: sexual selection by cryptic female choice.* Princeton, NJ: Princeton University Press.

Ehrhardt, A. A. & Baker, S. W. (1974). Fetal androgens, human central nervous system differentiation, and behavior sex differences. In *Sex differences in behavior,* ed. R. C. Friedman, pp. 33–51. New York: Wiley.

Eisenberg, J. F. (1981). *The mammalian radiations: an analysis of trends in evolution, adaptation, and behavior.* Chicago: University of Chicago Press.

Ekman, P., Friesen, W. V., O'Sullivan, M., Chan, A., Diacoyanni-Tarlatzis, I., Heider, K., Krause, R., LeCompte, W. A., Pitcairn, T. & Ricci-Bitti, P. E. (1987). Universals and cultural differences in the judgments of facial expressions of emotion. *Journal of Personality and Social Psychology,* **53**, 712–17.

Emlen, S. T. & Oring, L. W. (1977). Ecology, sexual selection, and the evolution of mating systems. *Science,* **197**, 215–23.

Eng, R., Gold, R. M. & Sawchenko, P. E. (1978). Hypothalamic hypoactivity prevented but not reverse by subdiaphragmatic vagotomy. *Physiology and Behaviour,* **20**, 637–41.

Epstein, A. N. (1978). Consensus, controversies, and curiosities. *Federations Proceedings,* **37**, 2711–16.

Epstein, A. N. & Stellar, E. (1955). The control of salt preference in the adrenalectomized rat. *Journal of Comparative and Physiological Psychology,* **48**, 167–72.

Erickson, C. J. & Zenone, P. G. (1976). Courtship differences in male ring doves: avoidance of cuckoldry? *Science,* **192**, 1353–4.

Ericsson, R. J., Langevin, C. N. & Nishino, M. (1973). Isolation of fractions rich in human Y sperm. *Nature,* **246**, 421–4.

Ernits, T. & Corbit, J. D. (1973). Taste as a dipsogenic stimulus. *Journal of Comparative and Physiological Psychology,* **83**, 27–31.

Falk, J. F. (1961). Production of polydipsia in normal rats. *Science,* **133**, 195–6.

Falk, J. F. (1972). The nature and determinants of adjunctive behavior. In *Schedule effects: drugs, drinking, and aggression,* ed. R. M. Gilbert & J. D. Keehn, pp. 148–73. Toronto: University of Toronto Press.

Falk, J. L. (1969). Conditions producing psychogenic polydipsia in animals. *Annals of the New York Academy of Sciences,* **157**, 569–93.

Falk, J. L. (1977). The origins and functions of adjunctive behaviour. *Animal Learning and Behavior,* **5**, 325–35.

Fallon, A. E. & Rozin, P. (1985). Sex differences in perceptions of desirable body shape. *Journal of Abnormal Psychology,* **94**, 102–5.

Feierman, J. R. (1987). The ethology of psychiatric populations: an introduction. *Ethology and Sociobiology,* **8**, 1–8.

Feigley, D. A., Parsons, P. J., Hamilton, L. W. & Spear, N. E. (1972). Development of habituation to novel environments in the rat. *Journal of Comparative and Physiological Psychology,* **79**, 443–52.

Feldon, J. & Gray, J. A. (1981). The partial reinforcement extinction effect after treatment with chlordiazepoxide. *Psychopharmacology,* **73**, 269–75.

Felix, D. & Akert, K. (1974). The effect of angiotensin II on neurones of the cat subfornical organ. *Brain Research,* **76**, 350–3.

Ferguson, N. B. & Keesey, R. E. (1975). Effect of a quinine-adulterated diet upon

body weight maintenance in male rats with ventromedial hypothalamic lesions. *Journal of Comparative and Physiological Psychology*, **89**, 478–88.
Feshbach, S. & Fraczek, A. (1979). *Aggression and behavior change: biological and social processes.* New York: Praeger.
Fiske, D. W. & Maddi, S. R. (1961). *Functions of varied experience.* Homewood, IL: Dorsey Press.
Fitch, R. H., Cowell, P. E., Schrott, L. M. & Denenberg, V. H. (1991). Corpus callosum: ovarian hormones and feminization. *Brain Research*, **542**, 313–17.
Fitch, R. H. & Denenberg, V. H. (1998). A role for ovarian hormones in sexual differentiation of the brain. *Behavioral and Brain Sciences*, **21**, 311–27.
Fitzsimons, J. T. (1961). Drinking by nephrectomized rats injected with various substances. *Journal of Physiology*, **155**, 563–79.
Fitzsimons, J. T. (1969). The role of a renal thirst factor in drinking induced by extracellular stimuli. *Journal of Physiology (London)*, **201**, 349–68.
Fitzsimons, J. T. (1972). Thirst. *Physiology Review*, **52**, 468–561.
Fitzsimons, J. T. (1978). Angiotensin, thirst, and sodium appetite: retrospect and prospect. *Federations Proceedings*, **37**, 2669–75.
Fitzsimons, J. T. & Kucharczyk, J. (1978). Drinking and haemodynamic changes induced in the dog by intracranial injection of components of the renin-angiotensin system. *Journal of Physiology (London)*, **276**, 419–34.
Fitzsimons, J. T. & Le Magnen, J. (1969). Eating as a regulatory control of drinking. *Journal of Comparative and Physiological Psychology*, **67**, 273–83.
Fitzsimons, J. T. & Simons, B. J. (1969). The effect on drinking in the rat of intravenous infusion of angiotensin, given alone or in combination with other stimuli of thirst. *Journal of Physiology (London)*, **203**, 45–57.
Fleming, A. (1986). Psychobiology of rat maternal behavior: how and where hormones act to promote maternal behavior at parturition. *Annals of the New York Academy of Sciences*, **474**, 234–51.
Fleming, A. S. (1989). Maternal responsiveness in human and animal mothers. *New Directions for Child Development*, **43**, 31–47.
Fleming, A. S., Corter, C. & Steiner, M. (1995). Sensory and hormonal control of maternal behavior in rat and human mothers. In *Motherhood in human and nonhuman primates: biosocial determinants*, ed. C. R. Pryce, R. D. Martin & D. Skuse, pp. 106–14. Basel: S. Karger, AG.
Fleming, A. S., Korsmit, M. & Deller, M. (1994). Rat pups are potent reinforcers to the maternal animal: effects of experience, parity, hormones, and dopamine function. *Psychobiology*, **22**, 44–53.
Fleming, A. S. & Rosenblatt, J. S. (1974). Maternal behavior in the virgin and lactating rat. *Journal of Comparative and Physiological Psychology*, **86**, 957–72.
Fleming, A. S., Steiner, M. & Anderson, V. (1987). Hormonal and attitudinal correlates of maternal behaviour during the early postpartum period in first-time mothers. *Journal of Reproductive and Infant Psychology*, **5**, 193–205.
Fleming, A. S., Steiner, M. & Corter, C. (1997). Cortisol, hedonics, and maternal responsiveness in human mothers. *Hormones and Behavior*, **32**, 85–98.
Fletcher, P. J. & Davies, M. (1990). Effects of 8-OH-DPAT, buspirone and ICS 205-930 on feeding in a novel environment: comparisons with chlordiazepoxide and FG 7142. *Psychopharmacology*, **102**, 301–8.
Fraczek, A. (1992). Patterns of aggressive-hostile behavior orientation among adolescent boys and girls. In *Of mice and women: aspects of female aggression*, ed. K. Bjorkqvist & P. Niemela, pp. 107–12. San Diego: Academic Press.
Franchina, J. J. & Dyer, A. B. (1989). Cross-modality transfer effects in conditioning-enhanced neophobia in chicks (*Gallus domesticus*): evidence for the separability of novelty from specific stimulus characteristics. *Animal Learning and*

Behavior, **17**, 261–5.

Frank, R. A. & Raudenbush, B. (1996). Individual differences in approach to novelty; the case for human neophobia. In *Viewing psychology as a whole*, ed. R. R. Hoffman, M. F. Sherrick & J. S. Warm, pp. 227–45. Washington, DC: American Psychological Association.

Frank, R. H. (1994). Group selection and 'geniuine' altruism. *Behavioral and Brain Sciences*, **17**, 620.

Franken, R. E. (1994). *Human motivation*. Pacific Grove, CA: Brooks/Cole.

Franklin, K. B. & Herberg, L. J. (1974). Ventromedial syndrome: the rat's 'finickiness' results from the obesity, not from the lesions. *Journal of Comparative and Physiological Psychology*, **87**, 410–14.

Fregly, M. J. & Waters, I. W. (1966). Effect of mineralcorticoids on spontaneous NaCl intake of adrenalectomized rats. *Hormones and Behavior*, **1**, 65–74.

Freud, S. & Riviere, J. (1930). *Civilization and its discontents*. New York: Jonathan Cape.

Friedman, M. I. & Stricker, E. M. (1976). The physiological psychology of hunger: a physiological perspective. *Psychological Review*, **83**, 409–31.

Frisch, R. E. (1987). Body fat, menarche, fitness and fertility. *Human Reproduction*, **2**, 521–33.

Frisch, R. E. & McArthur, J. W. (1974). Menstrual cycles: fatness as a determinant of minimum weight for height necessary for their maintenance or onset. *Science*, **185**, 949–51.

Frodi, A. M. & Lamb, M. E. (1978). Sex differences in responsiveness to infants: a developmental study of psychophysiological and behavioral responses. *Child Development*, **49**, 1182–8.

Gaffen, E. A. & Davies, J. (1982). Reward, novelty and spontaneous alternation. *Quarterly Journal of Experimental Psychology*, **34B**, 31–47.

Galef, B. G., Jr. (1977). Mechanisms for the transmission of acquired patterns of feeding from adult to juvenile rats. In *Learning mechanisms in food selection*, ed. L. M. Barker, M. R. Best & M. Domjan, pp. 123–48. Waco, TX: Baylor University Press.

Galef, B. G., Jr. & Beck, M. (1990). Diet selection and poison avoidance by mammals individually and in social groups. In *Neurobiology of food and fluid intake. Handbook of behavioral neurobiology*, ed. E. M. Stricker, pp. 329–49. New York: Plenum Press.

Gangestad, S. W. & Buss, D. M. (1993). Pathogen prevalence and human mate preferences. *Ethology and Sociobiology*, **14**, 89–96.

Garcia, J., Ervin, F. R., Yorke, C. H. & Koelling, R. A. (1967). Conditioning with delayed vitamin injections. *Science*, **155**, 716–18.

Garcia, J. & Koelling, R. A. (1966). Relation of cue to consequence. *Psychonomic Science*, **4**, 123–4.

Garfinkel, P. E. & Garner, D. M. (1982). *Anorexia nervosa: a multidimensional perspective*. New York: Brunner/Mazel.

Garn, S. M. & Leonard, W. R. (1989). What did our ancestors eat? *Nutrition Reviews*, **47**, 337–45.

Garrow, J. S., Durrant, M. L., Mann, S., Stalley, S. F. & Warwick, P. M. (1978). Factors determining weight loss in obese patients in a metabolic ward. *International Journal of Obesity and Related Dysfunctions*, **2**, 441–7.

Geen, R. G. (1995). *Human motivation: a social psychological approach*. Pacific Grove, CA: Brooks/Cole.

Gibber, J. R. (1986). Infant-directed behavior of rhesus monkeys during their first pregnancy and parturition. *Folia Primatologica*, **46**, 118–24.

Gibber, J. R. & Goy, R. W. (1985). Infant-directed behavior in young rhesus monkeys: sex differences and effects of prenatal androgens. *American Journal of*

Primatology, **8**, 225–37.

Gibbs, J., Falasco, J. D. & McHugh, P. R. (1976). Cholecystokinin-decreased food intake in rhesus monkeys. *American Journal of Physiology*, **230**, 15–18.

Gibbs, J., Maddison, S. P. & Rolls, E. T. (1981). Satiety role of the small intestine examined in sham-feeding rhesus monkeys. *Journal of Comparative and Physiological Psychology*, **95**, 1003–15.

Gibbs, J., Young, R. C. & Smith, G. P. (1973). Cholecystokinin decreases food intake in rats. *Journal of Comparative and Physiological Psychology*, **84**, 488–95.

Gilbert, R. M. & Keehn, J. D. (1972). *Schedule effects: drugs, drinking and aggression.* Toronto: University of Toronto Press.

Gilman, A. (1937). The relation between blood osmotic pressure, fluid distribution, and voluntary intake. *American Journal of Physiology*, **120**, 323–8.

Glow, P. H. (1970). Some acquisition and performance characteristics of response contingent reinforcement. *Australian Journal of Psychology*, **22**, 145–54.

Gold, R. M., Jones, A. P. & Sawchenko, P. E. (1977). Paraventricular area: critical focus of a longitudinal neurocircuitry mediating food intake. *Physiology and Behavior*, **18**, 1111–19.

Goldfoot, D. A., Wallen, K., Neff, D. A., McBrair, M. C. & Goy, R. W. (1984). Social influence on the display of sexually dimorphic behavior in rhesus monkeys: isosexual rearing. *Archives of Sexual Behavior*, **13**, 395–412.

Goodall, J. (1986). *The behaviour of free-living chimpanzees in the Gombe Stream reserve.* London: Ballière Tindall & Cassell.

Goodfellow, P. N. & Lovell-Badge, R. (1993). SRY and sex determination in mammals. *Annual Review of Genetics*, **27**, 71–92.

Gorski, R. A., Gordon, J. H., Shryne, J. E. & Southam, A. M. (1978). Evidence for a morphological sex difference within the medial preoptic area of the rat brain. *Brain Research*, **148**, 333–46.

Gould, J. L. & Gould, C. G. (1997). *Sexual selection: mate choice and courtship in nature.* New York: Scientific American Library.

Gould, S. J. & Lewontin, R. (1979). The spandrels of San Marco and the Panglossian paradigm: a critique of the adaptationist programme. *Proceedings of the Royal Society of London (B)*, **205**, 581–98.

Goy, R. W. (1970). Experimental control of psychosexuality. *Philosophical Transactions of the Royal Society of London. Series B: Biological Sciences*, **259**, 149–62.

Graff, H. & Stellar, E. (1962). Hyperphagia, obesity and finickiness. *Journal of Comparative and Physiological Psychology*, **55**, 418–24.

Grandison, L. & Guidotti, A. (1977). Stimulation of food intake by muscimol and beta endorphin. *Neuropharmacology*, **16**, 533–6.

Gray, J. (1976). The behavioural inhibition system: a possible substrate for anxiety. In *Theoretical and experimental bases of behavioural modification*, ed. M. P. Feldman & A. M. Broadhurst, pp. 3–41. London: Wiley.

Gray, J. A. (1969). Sodium amobarbital and effects of frustrative nonreward. *Journal of Comparative and Physiological Psychology*, **69**, 55–64.

Gray, J. A. (1970). Sodium amobarbital, the hippocampal theta rhythm, and the partial reinforcement extinction effect. *Psychological Review*, **77**, 465–80.

Gray, J. A. (1971). *The psychology of fear and stress.* New York: McGraw-Hill.

Gray, J. A. (1982). *The neuropsychology of anxiety: an enquiry into the functions of the septo-hippocampal system.* New York: Clarendon Press/Oxford University Press.

Gray, J. A. (1987). *The psychology of fear and stress.* Cambridge: Cambridge University Press.

Gray, J. A. & Ball, G. G. (1970). Frequency-specific relation between hippocampal theta rhythm, behavior, and amobarbital action. *Science*, **168**, 1246–8.

Gray, J. A., Lean, J. & Keynes, A. (1969). Infant androgen treatment and adult open-field behavior: direct effects and effects of injection to siblings. *Physiology and Behavior,* **4**, 178–81.

Green, S. (1987). *Physiological psychology: an introduction.* London: Routledge & Kegan Paul.

Grill, H. J. & Berridge, K. C. (1985). Taste reactivity as a measure of the neural control of palatability. In *Progress in psychobiology and physiological psychology,* vol. II, ed. J. M. Sprague & A. N. Epstein. New York: Academic Press.

Grill, H. J. & Norgren, R. (1978). The taste reactivity test. I. Mimetic responses to gustatory stimuli in neurologically normal rats. *Brain Research,* **143**, 263–79.

Grossman, K., Thane, K. & Grossman, K. E. (1981). Maternal tactual contact of the newborn after various conditions of mother–infant interaction. *Developmental Psychology,* **17**, 158–69.

Grossman, S. P. (1960). Eating or drinking elicited by direct adrenergic or cholinergic stimulation of hypothalamic mechanisms. *Science,* **132**, 301–2.

Grossman, S. P. (1966). The VMH: a center for affective reactions, satiety, or both? *Physiology and Behavior,* **1**, 1–10.

Grossman, S. P. (1967). *A textbook of physiological psychology.* New York: Wiley.

Grossman, S. P. (1972). Aggression, avoidance, and reaction to novel environments in female rats with ventromedial hypothalamic lesions. *Journal of Comparative and Physiological Psychology,* **78**, 274–83.

Gubernick, D. J. (1980). 'Maternal labelling' or maternal 'labelling' in goats? *Animal Behaviour,* **28**, 124–9.

Gubernick, D. J. (1981a). Parent and infant attachment in mammals. In *Parental care in mammals,* ed. D. J. Gubernick & P. H. Klopfer, pp. 243–305. New York: Plenum Press.

Gubernick, D. J. (1981b). Mechanisms of maternal 'labelling' in goats. *Animal Behaviour,* **29**, 305–6.

Gust, D. A., Gordon, T. P., Gergits, W. F., Casna, N. J., Gould, K. G. & M., M. H. (1996). Male dominance rank and offspring-initiated affiliative behaviours were not predictors of paternity in a captive group of pigtail macaques (*Macaca nemestrina*). *Primates,* **37**, 271–8.

Hall, J. F. (1961). *Psychology of motivation.* Chicago: Lippincott.

Halliday, M. S. (1966). Effect of previous exploratory activity on the exploration of a simple maze. *Nature,* **209**, 432–3.

Halliday, M. S. (1967). Exploratory behaviour in elevated and enclosed mazes. *Quarterly Journal of Experimental Psychology,* **19**, 254–63.

Halliday, T. & Slater, P. J. B. (1983). *Causes and effects.* Oxford: Blackwell Scientific.

Hallonquist, J. D. & Brandes, J. S. (1981a). Housing affects hyperreactivity but not obesity induced by medial hypothalamic lesions. *Physiology and Behavior,* **26**, 1025–9.

Hallonquist, J. D. & Brandes, J. S. (1981b). Ventromedial hypothalamic lesions and weight gain in rats: absence of a static phase. *Physiology and Behavior,* **27**, 709–13.

Hallonquist, J. D. & Brandes, J. S. (1984). Ventromedial hypothalamic lesions in rats: gradual elevation of body weight set-point. *Physiology and Behavior,* **33**, 831–6.

Halmi, K. A., Dekirmenjian, H., Davis, J. M., Casper, R. & Goldberg, S. (1978). Catecholamine metabolism in anorexia nervosa. *Archives of General Psychiatry,* **35**, 458–60.

Halperin, D. A. (1996). Cults and children: a group dynamic perspective on child abuse within cults. In *Group therapy with children and adolescents,* ed. P. Kymissis & D. A. Halperin, pp. 353–66. Washington, DC: American Psychiatric Press.

Haltmeyer, G. C., Denenberg, V. H. & Zarrow, M. X. (1967). Modification of the plasma corticosterone reponse as a function of infantile stimulation and electric shock parameters. *Physiology and Behavior,* **2**, 61–3.

Hamilton, W. D. (1964a). The genetical evolution of social behaviour. I. *Journal of Theoretical Biology*, **7**, 1–16.
Hamilton, W. D. (1964b). The genetical evolution of social behaviour. II. *Journal of Theoretical Biology*, **7**, 17–52.
Hamilton, W. D., Axelrod, R. & Tanese, R. (1990). Sexual reproduction as an adaptation to resist parasites (a review). *Proceedings of the National Academy of Sciences USA*, **87**, 3566–73.
Hamilton, W. D. & Zuk, M. (1982). Heritable true fitness and bright birds: a role for parasites? *Science*, **218**, 384–7.
Hammer, R. P., Jr. & Bridges, R. S. (1987). Preoptic area opioids and opiate receptors increase during pregnancy and decrease during lactation. *Brain Research*, **420**, 48–56.
Han, P. W. (1967). Hypothalamic obesity in rats without hyperphagia. *Transactions of the New York Academy of Science*, **30**, 229–43.
Hansen, E. W. (1971). Responsiveness of ring dove foster parents to squabs. *Journal of Comparative and Physiological Psychology*, **77**, 382–7.
Harlow, H. F. & Mears, C. (1979). *The human model: primate perspectives.* Washington, DC: Winston.
Harper, A. E. (1967). Effect of dietary protein content and amino acid pattern on food intake and preference. In *Handbook of physiology*. Section 6. *Alimentary canal*, vol. 1. *Control of food and water intake*, ed. C. F. Code, pp. 399–410. Washington, DC: American Physiological Association.
Harris, G. W. & Levine, S. (1965). Sexual differentiation of the brain and its experimental control. *Journal of Physiology (London)*, **181**, 379–400.
Harris, L. J., Clay, J., Hargreaves, F. J. & Ward, A. (1933). Appetite and choice of diet: the ability of the vitamin B-deficient rat to discriminate between diets containing and lacking the vitamin. *Proceedings of the Royal Society (London)*, **113**, 161–90.
Harris, M. (1987). *The sacred cow and the abominable pig: riddles of food and culture.* New York: Simon & Schuster.
Hebb, D. O. (1972). *A textbook of psychology*. Philadelphia: Saunders.
Heinsbroek, R. P., van Haaren, F. & van de Poll, N. E. (1988). Sex differences in passive avoidance behavior of rats: sex-dependent susceptibility to shock-induced behavioral depression. *Physiology and Behavior*, **43**, 201–6.
Hennessy, M. B., Hershberger, W. A., Bell, R. W. & Zachman, T. A. (1976). The influence of early auditory experience on later auditory and tactual variation seeking in the rat. *Developmental Psychobiology*, **9**, 255–60.
Hennessy, M. B., Smotherman, W. P. & Levine, S. (1977). Early olfactory enrichment enhances later consumption of novel substances. *Physiology and Behavior*, **19**, 481–3.
Hersher, L., Richmond, J. B. & Moore, A. U. (1963). Maternal behavior in sheep and goats. In *Maternal behavior in mammals*, ed. H. L. Rheingold, pp. 203–32. New York: Wiley.
Hess, C. & Blozovski, D. (1987). Hippocampal muscarinic cholinergic mediation of spontaneous alternation and fear in the developing rat. *Behavioural Brain Research*, **24**, 203–14.
Hewlet, B. S. (1987). Intimate fathers: patterns of holding among Aka pygmies. In *The father's role: cross-cultural perspectives*, ed. M. E. Lamb, pp. 295–330. Hillsdale, NJ: Lawrence Erlbaum Associates.
Hines, M. & Kaufman, F. R. (1994). Androgen and the development of human sex-typical behavior: rough-and-tumble play and sex of preferred playmates in children with congenital adrenal hyperplasia (CAH). *Child Development*, **65**, 1042–53.

Hoebel, B. G. & Teitelbaum, P. (1966). Weight regulation in normal and hypothalamic hyperphagic rats. *Journal of Comparative and Physiological Psychology*, **61**, 189–93.

Hofer, M. A. (1987). Early social relationships: a psychobiologist's view. *Child Development*, **58**, 633–47.

Hoffman, K. A., Mendoza, S. P., Hennessy, M. B. & Mason, W. A. (1995). Responses of infant titi monkeys, *Callicebus moloch*, to removal of one or both parents: evidence for paternal attachment. *Developmental Psychobiology*, **28**, 399–407.

Hollenberg, N. K. (1980). Set point for sodium homeostasis: surfeit, deficit, and their implications. *Kidney International*, **17**, 423–9.

Holmes, W. G. & Sherman, P. W. (1982). The ontogeny of kin recognition in two species of ground squirrels. *American Zoologist*, **22**, 491–517.

Holt, L. & Gray, J. A. (1983). Septal driving of the hippocampal theta rhythm produces a long-term, proactive and non-associative increase in resistance to extinction. *Quarterly Journal of Experimental Psychology*, **35B**, 97–118.

Hudson, R. & Distel, H. (1982). The pattern of behaviour of rabbit pups in the nest. *Behaviour*, **79**, 255–71.

Hughes, R. N. (1997). Spontaneous alternation in animals: mechanisms, motives, and application. In *Viewing psychology as a whole*, ed. R. R. Hoffman, M. F. Sherrick & J. S. Warm, pp. 269–86. Washington, DC: American Psychological Association.

Hunt, H. F. (1963). Early 'experience' and its effects of later behavioral processes in rats. I. Initial experiments. *Transactions of the New York Academy of Science*, **25**, 858–70.

Hunt, H. F. & Otis, L. S. (1955). Restricted experience and 'timidity' in the rat. *American Psychologist*, **10**, 432.

Hustvedt, B. E. & Lovo, A. (1972). Correlation between hyperinsulinemia and hyperphagia in rats with ventromedial hypothalamic lesions. *Acta Physiologica Scandinavica*, **84**, 29–33.

Hwang, C. P. (1987). Cesarean childbirth in Sweden: effects on the mother and father–infant relationship. *Infant Mental Health Journal*, **8**, 91–9.

Imperato-McGinley, J., Guerrero, L., Gautier, T. & Peterson, R. E. (1974). Steroid 5alpha-reductase deficiency in man: an inherited form of male pseudohermaphroditism. *Science*, **186**, 1213–15.

Ingram, J. C. (1977). Interactions between parents and infants and the development of independence in the common marmoset. *Animal Behaviour*, **25**, 811–27.

Insel, T. R. (1990). Oxytocin and maternal behavior. In *Mammalian parenting: biochemical, neurobiological and behavioral determinants*, ed. N. A. Krasnegor & R. S. Bridges, pp. 260–80. New York: Oxford University Press.

Insel, T. R., Young, L. & Wang, Z. (1997). Central oxytocin and reproductive behaviours. *Review of Reproduction*, **2**, 28–37.

Irons, W. (1991). Anthropology. In *The sociobiological imagination*, ed. M. Mary, pp. 71–90. Albany, NY: State University of New York Press.

Ison, J. R. & Pennes, E. S. (1969). Interaction of amobarbital sodium and reinforcement schedule in determining resistance to extinction of an instrumental running response. *Journal of Comparative and Physiological Psychology*, **68**, 215–19.

Jacobs, K. M., Marks, G. P. & Scott, T. R. (1988). Taste responses in the nucleus traxtus solitarius of sodium-deficient rats. *Journal of Physiology*, **406**, 393–410.

Janzen, D. H. (1977). Why fruits rot, seeds mold and meats spoil. *American Naturalist*, **111**, 691–713.

Johnson, A. K. & Buggy, J. (1978). Periventricular preoptic-hypothalamus is vital for

thirst and normal water economy. *American Journal of Physiology*, **234**, R122–9.
Jones, R. L. & Hanson, H. C. (1985). *Mineral licks, geophagy, and biogeochemistry of North American ungulates.* Ames: Iowa State University Press.
Juraska, J. M. (1991). Sex differences in 'cognitive' regions of the rat brain. Special Issue: neuroendocrine effects on brain development and cognition. *Psychoneuroendocrinology*, **16**, 105–19.
Katz, J. L. & Weiner, H. (1975). Editorial: a functional, anterior hypothalamic defect in primary anorexia nervosa? *Psychosomatic Medicine*, **37**, 103–5.
Katz, R. J. (1988). Endorphins, exploration and activity. In *Endorphins, opiates and behavioural processes.*, ed. R. J. Rodgers & S. J. Cooper, pp. 249–67. Chichester: Wiley.
Katz, S. H. (1982). Food, behavior, and biocultural evolution. In *The psychobiology of human food selection*, ed. L. M. Barker, pp. 171–88. Westport, CN: AVI.
Keesey, R. E., Corbett, S. W., Hirvonen, M. D. & Kaufman, L. N. (1984). Heat production and body weight changes following lateral hypothalamic lesions. *Physiology and Behavior*, **32**, 309–17.
Keesey, R. E. & Hirvonen, M. D. (1997). Body weight set-points: determination and adjustment. *Journal of Nutrition*, **127**, 1875S-83S.
Keesey, R. E. & Powley, T. L. (1975). Hypothalamic regulation of body weight. *American Scientist*, **63**, 558–65.
Kennedy, J. S. (1992). *The new anthropomorphism.* Cambridge: Cambridge University Press.
Kenrick, D. T. & Keefe, R. C. (1992). Age preferences in mates reflect sex differences in human reproductive strategies. *Behavioral and Brain Sciences*, **15**, 75–133.
Kesner, R. P., Berman, R. F. & Tardif, R. (1992). Place and taste aversion learning: role of basal forebrain, parietal cortex, and amygdala. *Brain Research Bulletin*, **29**, 345–53.
Keverne, E. B. (1988). Central mechanisms underlying the neural and neuroendocrine determinants of maternal behaviour. *Psychoneuroendocrinology*, **13**, 127–41.
Keverne, E. B. (1995). Neurochemical changes accompanying the reproductive process: their significance for maternal care in primates and other mammals. In *Motherhood in human and nonhuman primates: biosocial determinants*, ed. C. R. Pryce, R. D. Martin & D. Skuse, pp. 69–77. Basel: S. Karger.
Kimble, D. P. (1968). Hippocampus and internal inhibition. *Psychological Bulletin*, **70**, 285–95.
King, B. M. & Frohman, L. A. (1982). The role of vagally-mediated hyperinsulinemia in hypothalamic obesity. *Neuroscience and Biobehavioral Reviews*, **6**, 205–14.
Kirchgessner, A. L. & Sclafani, A. (1988). PVN-hindbrain pathway involved in the hypothalamic hyperphagia–obesity syndrome. *Physiology and Behavior*, **42**, 517–28.
Kirkby, R. J., Stein, D. G., Kimble, R. J. & Kimble, D. P. (1967). Effects of hippocampal lesions and duration of sensory input on spontaneous alternation. *Journal of Comparative and Physiological Psychology*, **64**, 342–5.
Kleiman, D. G. & Malcolm, J. R. (1981). The evolution of male parental investment in mammals. In *Parental care in mammals*, ed. D. G. Gubernick & P. H. Klopfer, pp. 347–87. New York: Plenum Press.
Klopfer, P. H., Adams, D. K. & Klopfer, M. S. (1964). Maternal 'imprinting' in goats. *Proceeding of the National Academy of Sciences USA*. **52**, 911–14.
Klopfer, P. H. & Gamble, J. (1966). Maternal 'imprinting' in goats: the role of chemical senses. *Zeitschrift für Tierpsychologie*, **23**, 588–92.
Kodric-Brown, A. & Brown, J. H. (1984). Truth in advertising: the kinds of traits favored by sexual selection. *American Naturalist*, **127**, 309–23.
Koivisto, U. K. & Sjoden, P. O. (1996). Food and general neophobia in Swedish

families: parent–child comparisons and relationships with serving specific foods. *Appetite,* **26**, 107–18.

Kolakowska, L., Larue-Achagiotis, C. & Le Magnen, J. (1984). Comparative effects of lesion of the basolateral nucleus and lateral nucleus of the amygdaloid body on neophobia and conditioned taste aversion in the rat. *Physiology and Behavior,* **32**, 647–51.

Kraemer, G. W. (1992). A psychobiological theory of attachment. *Behavioral and Brain Sciences,* **15**, 493–541.

Kraemer, G. W. (1995). Significance of social attachment in primate infants: the infant–caregiver relationship and volition. In *Motherhood in human and nonhuman primates: biosocial determinants.,* ed. C. R. Pryce, R. D. Martin & D. Skuse, pp. 152–61. Basel: S. Karger, AG.

Kraly, F. S. (1983). Histamine plays a part in induction of drinking by food intake. *Nature,* **302**, 65–6.

Kraly, F. S. (1990). Drinking elicited by eating. In *Progress in psychobiology and physiological psychology,* ed. A. N. Epstein & A. R. Morrison, pp. 64–133. New York: Academic Press.

Kraly, F. S. & Corneilson, R. (1990). Angiotensin II mediates drinking elicited by eating in the rat. *American Journal of Physiology,* **258**, R436–42.

Kraly, F. S., Tribuzio, R. A., Keefe, M. E., Kim, Y. M. & Lowrance, R. (1995). Endogenous histamine contributes to drinking initiated without postprandial challenges to fluid homeostasis in rats. *Physiology and Behavior,* **58**, 1137–43.

Krieckhaus, E. E. (1970). 'Innate recognition' aids rats in sodium regulation. *Journal of Comparative and Physiological Psychology,* **73**, 117–22.

Krieckhaus, E. E. & Wolf, G. (1971). Acquisition of sodium by rats: interaction of innate mechanisms and latent learning. *Journal of Comparative and Physiological Psychology,* **65**, 197–201.

Kucharczyk, J., Assaf, S. Y. & Mogenson, G. J. (1976). Differential effects of brain lesions on thirst induced by the administration of angiotensin-II to the preoptic region, subfornical organ and anterior third ventricle. *Brain Research,* **108**, 327–37.

Kurland, J. A. (1979). Paternity, mother's brother, and sociality. In *Evolutionary biology and human social behavior: an anthropological perspective,* ed. N. A. Chagnon, pp. 145–80. North Scituate, MA: Duxbury Press.

Kyrkouli, S. E., Stanley, B. G., Hutchison, R., Seirafi, R. D. & Leibowitz, S. F. (1990). Peptide-amine interactions in the hypothalamic paraventricular nucleus: analysis of galanin and neuropeptide Y in relation to feeding. *Brain Research,* **521**, 185–91.

Lack, D. L. (1968). *Ecological adaptations for breeding in birds.* London: Methuen.

Lahari, R. K. & Southwick, C. H. (1966). Paternal care in *Macaca sylvana. Folia Primatologica,* **4**, 267–14.

Lamas, E. & Pellon, R. (1997). Food deprivation and food-delay effects on the development of adjunctive drinking. *Physiology and Behavior,* **61**, 153–18.

Larsson, K. (1994). The psychobiology of parenting in mammals. *Scandinavian Journal of Psychology,* **35**, 97–143.

Lat, J. (1967). Self-selection of dietary components. In *Handbook of physiology.* Section 6. *Alimentary canal,* ed. C. F. Code, pp. 367–86. Washington DC: American Physiological Association.

Lazar, J. W. (1974). A comparison of some theoretical proposals of J. R. Kantor and T.C. Schneirla. *Psychological Record,* **24**, 177–90.

Le Boeuf, B. J. (1972). Sexual behaviour in the northern elephant seal. *Behaviour,* **41**, 1–26.

Le Magnen, J. (1992). *Neurobiology of feeding and nutrition.* San Diego: Academic Press.

Le Magnen, J. (1998). Synthetic approach to the neurobiology of behaviour. *Appetite,* **31**, 1–8.

Lee, E. H., Teng, Y. P. & Chai, C. Y. (1987). Stress and corticotropin-releasing factors potentiate center region activity of mice in an open field. *Neuroendocrinology,* **93**, 320–3.

Lee, M. C., Thrasher, T. N. & Ramsay, D. J. (1981). Is angiotensin essential in drinking induced by water deprivation and caval ligation? *American Journal of Physiology,* **240**, R75–80.

Lehrman, D. S. (1955). The physiological basis of parental behavior in the ring dove. *Behaviour,* **7**, 241–86.

Lehrman, D. S. (1964). The reproductive behavior of ring doves. *Scientific American,* **211**, 48–54.

Lehrman, D. S. (1971). Experiential background for the induction of reproductive behavior. In *Biopsychology of development,* ed. E. Tobach, pp. 297–302. New York: Academic Press.

Lehrman, D. S. & Wortis, R. P. (1967). Breeding experience and breeding efficiency in the ring dove. *Animal Behaviour,* **15**, 223–8.

Leibowitz, S. F. (1978). Paraventricular nucleus: a primary site mediating adrenergic stimulation of feeding and drinking. *Pharmacology, Biochemistry and Behavior,* **8**, 163–75.

Leibowitz, S. F. (1986). Brain monoamines and peptides: role in the control of eating behavior. *Federations Proceedings,* **45**, 1396–403.

Leibowitz, S. F. & Brown, L. L. (1980). Histochemical and pharmacological analysis of noradrenergic projections to the paraventricular hypothalamus in relation to feeding stimulation. *Brain Research,* **201**, 289–314.

Leibowitz, S. F., Weiss, G. F. & Shor-Posner, G. (1988). Hypothalamic serotonin: pharmacological, biochemical, and behavioral analyses of its feeding-suppressive action. *Clinical Neuropharmacology,* **11**, S51–71.

Lenhardt, M. L. (1977). Vocal contour cues in maternal recognition in goat kids. *Applied Animal Ethology,* **3**, 211–19.

Leon, M. & Moltz, H. (1972). The development of the pheromonal bond in the albino rat. *Physiology and Behavior,* **8**, 683–6.

Lester, D. (1967). Exploratory response of rats to stimuli of differing complexity. *Perceptual and Motor Skills,* **24**, 1333–4.

Levine, R. & Levine, S. (1989). Role of the pituitary–adrenal hormones in the acquisition of schedule- induced polydipsia. *Behavioral Neuroscience,* **103**, 621–37.

Levine, S. (1969). Infantile stimulation: a perspective. In *Stimulation in early infancy,* ed. A. Ambrose, pp. 3–19. London: Academic Press.

Levine, S., Coe, C. L., Smotherman, W. P. & Kaplan, J. (1978). Prolonged cortisol elevation in the infant squirrel monkey after reunion with mother. *Physiology and Behavior,* **20**, 7–10.

Levine, S., Haltmeyer, G. C., Karas, G. G. & Denenberg, V. H. (1967). Physiological and behavioral effects of infantile stimulation. *Physiology and Behavior,* **2**, 55–9.

Levine, S. & Lewis, G. W. (1959). Critical periods for the effects of infantile experience on maturation of the stress response. *Science,* **129**, 42–3.

Levine, S. & Mullins, R. F. (1966). Hormonal influences on brain organization in infant rats. *Science,* **152**, 1585–92.

Levine, S. & Wiener, S. G. (1988). Psychoendocrine aspects of mother–infant relationships in nonhuman primates. *Psychoneuroendocrinology,* **13**, 143–54.

Levitsky, D. & Collier, G. (1968). Schedule-induced wheel-running. *Physiology and Behavior,* **3**, 571–3.

Lévy, F. (1985). Contribution à l'analyse des mécanismes de mise en place du comportment maternal chez la brébis (*Ovis aries*): étude de la répulsion et de l'attraction vis-à-vis du liquide amniotique, mis en evidence, determinisme, role. Thèse de Doctorat de l'Université Paris VI, Paris.

Lévy, F. & Poindron, P. (1987). The importance of amniotic fluids for the establishment of maternal behaviour in experienced and inexperienced ewes. *Animal Behaviour*, **35**, 687–92.

Lewis, C. (1986). *Becoming a father.* Milton Keynes: Open University Press.

Leyton, E. & Locke, G. (1998). *Touched by fire: Doctors without Borders in a third world crisis.* Toronto: McClelland and Stewart.

Li, B. H. & Rowland, N. E. (1994). Cholecystokinin- and dexfenfluramine-induced anorexia compared using devazepide and c-Fos expression in the rat brain. *Regulatory Peptides*, **50**, 223–33.

Li, B. H., Spector, A. C. & Rowland, N. E. (1994). Reversal of dexfenfluramine-induced anorexia and c-Fos/c-Jun expression by lesion in the lateral parabrachial nucleus. *Brain Research*, **640**, 255–67.

Lisk, R. D. & Suydam, A. J. (1967). Sexual behavior patterns in the prepubertally castrate rat. *Anatomical Record*, **157**, 181–9.

Logue, A. W. (1991). *The psychology of eating and drinking: an introduction.* New York: W.H. Freeman.

Lorenz, K. (1971). *On aggression.* New York: Bantam Books.

Lovejoy, C. O. (1981). The origin of man. *Science*, **211**, 341–50.

Luby, E. D., Marrazzi, M. A. & Sperti, S. (1987). Anorexia nervosa: a syndrome of starvation dependence. *Comprehensive Therapy*, **13**, 16–21.

Lumsden, C. J. & Wilson, E. O. (1981). *Genes, mind, and culture: the coevolutionary process.* Cambridge, MA: Harvard University Press.

Lyon, B. E., Montgomerie, R. D. & Hamilton, L. D. (1987). Male parental care and monogamy in snow buntings. *Behavioral Ecology and Sociobiology*, **20**, 377–82.

MacLean, P. D. (1970). The limbic brain in relation to the psychoses. In *Physiological correlates of emotion*, ed. P. D. Black, pp. 129–46. New York: Academic Press.

MacLusky, N. J. & Naftolin, F. (1981). Sexual differentiation of the central nervous system. *Science*, **211**, 1294–302.

Maren, S., Tocco, G., Chavanne, F., Baudry, M., Thompson, R. F. & Mitchell, D. (1994). Emergence neophobia correlates with hippocampal and cortical glutamate receptor binding in rats. *Behavioral and Neural Biology*, **62**, 68–72.

Marrazzi, M. A., Kinzie, J. & Luby, E. D. (1995). A detailed longitudinal analysis on the use of naltrexone in the treatment of bulimia. *International Journal of Clinical Psychopharmacology*, **10**, 173–6.

Marrazzi, M. A., Mullings-Britton, J., Stack, L. & Powers, R. J. (1990). Atypical endogenous opioid systems in mice in relation to an auto-addiction opioid model of anorexia nervosa. *Life Sciences*, **47**, 1427–35.

Marshall, J. F. & Teitelbaum, P. (1974). Further analysis of sensory inattention following lateral hypothalamic damage in rats. *Journal of Comparative and Physiological Psychology*, **86**, 375–95.

Marshall, J. F., Turner, B. H. & Teitelbaum, P. (1971). Sensory neglect produced by lateral hypothalamic damage. *Science*, **174**, 523–5.

Maslow, A. H. (1954). *Motivation and personality.* New York: Harper.

Masserman, J. H. (1941). Is the hypothalamus a center of emotion? *Psychosomatic Medicine*, **2**, 1–25.

Maynard Smith, J. & Harper, D. G. (1988). The evolution of aggression: can selection generate variability? *Philosophical Transactions of the Royal Society of London. Series B: Biological Sciences*, **319**, 557–70.

McBurney, D. H. & Bartoshuk, L. M. (1973). Interactions between stimuli with different taste qualities. *Physiology and Behavior*, **10**, 1101–6.

McClelland, D. C. (1987). *Human motivation*. Cambridge: Cambridge University Press.

McClintock, M. K. (1998a). On the nature of mammalian and human pheromones. *Annals of the New York Academy of Sciences*, **855**, 390–2.

McClintock, M. K. (1998b). Whither menstrual synchrony? *Annual Review of Sex Research*, **9**, 77–95.

McEwen, B. S. (1976). Interactions between hormones and nerve tissue. *Scientific American*, **235**, 48–58.

McFarland, D. (1989). *Problems of animal behaviour*. Harlow: Longman Scientific and Technical.

McFarland, D. (1993). *Animal behaviour: psychobiology, ethology, and evolution*. Harlow: Longman Scientific and Technical.

McFarland, D. J. & Sibly, R. M. (1975). The behavioural final common path. *Philosophical Transactions of the Royal Society of London. Series B: Biological Sciences*, **270**, 265–93.

McGowan, M. K., Brown, B. & Grossman, S. P. (1988). Depletion of neurons from preoptic area impairs drinking to various dipsogens. *Physiology and Behavior*, **43**, 815–22.

McGrew, W. C. (1988). Parental division of infant caretaking varies with family composition in cotton-top tamarins. *Animal Behaviour*, **36**, 285–6.

McHugh, P. R. & Moran, T. H. (1986). The stomach, cholecystokinin, and satiety. *Federations Proceedings*, **45**, 1384–90.

McKinley, M. J., Denton, D. A. & Weisinger, R. S. (1978). Sensors for antidiuresis and thirst-osmoreceptors or CSF sodium detectors? *Brain Research*, **141**, 89–103.

McMichael, R. E. (1961). The effects of pre-weaning shock and gentling on later resistance to stress. *Journal of Comparative and Physiological Psychology*, **54**, 416–21.

Meaney, M. J., Aitken, D. H., Bodnoff, S. R., Iny, L. J., Tatarewicz, J. E. & Sapolsky, R. M. (1985). Early postnatal handling alters glucocorticoid receptor concentrations in selected brain regions. *Behavioral Neuroscience*, **99**, 765–70.

Meaney, M. J., Bhatnagar, S., Larocque, S., McCormick, C., Shanks, N., Sharma, S., Smythe, J., Viau, V. & Plotsky, P. M. (1993). Individual differences in the hypothalamic–pituitary–adrenal stress response and the hypothalamic CRF system. *Annals of the New York Academy of Science*, **697**, 70–85.

Meaney, M. J., Stewart, J., Poulin, P. & McEwen, B. S. (1983). Sexual differentiation of social play in rat pups is mediated by the neonatal androgen-receptor system. *Neuroendocrinology*, **37**, 85–90.

Mendoza, S. P., Coe, C. L., Smotherman, W. P., Kaplan, J. & Levine, S. (1980). Functional consequences of attachment: a comparison of two species. In *Maternal influences and early behavior*, ed. R. W. Bell & W. P. Smotherman, pp. 235–52. Jamaica, NY: Spectrum Publications.

Mendoza, S. P. & Mason, W. A. (1986). Parental division of labour and differentiation of attachments in a monogamous primate (*Callicebus moloch*). *Animal Behaviour*, **34**, 1336–47.

Menzel, E. W., Davenport, R. K. & Rogers, C. M. (1961). Some aspects of behavior toward novelty in young chimpanzees. *Journal of Comparative and Physiological Psychology*, **54**, 16–19.

Milinski, M. (1997). Tit for tat in sticklebacks and the evolution of cooperation. *Nature*, **325**, 433–5.

Miller, N. E. (1948). Studies in fear as an aquirable drive. I. Fear as motivation and fear-reduction as reinforcement in the learning of new responses. *Journal of Experimental Psychology*, **38**, 89–101.

Miller, N. E. (1959). Liberalization of basic S–R concepts: extensions to conflict behavior, motivation and social learning. In *Psychology; a study of a science*, vol.

2, ed. S. Koch, pp. 196–292. New York: McGraw-Hill.

Miller, R. R. & Holzman, A. D. (1981). Neophobia: generality and function. *Behavioral and Neural Biology*, **33**, 17–44.

Minor, T. R., Dess, N. K., Ben-David, E. & Chang, W. C. (1994). Individual differences in vulnerability to inescapable shock in rats. *Journal of Experimental Psychology: Animal Behavior Processes*, **20**, 402–12.

Miselis, R. R. (1981). The efferent projections of the subfornical organ of the rat: a circumventricular organ within a neural network subserving water balance. *Brain Research*, **230**, 1–23.

Mitchell, D. (1976). Experiments on neophobia in wild and laboratory rats: a reevaluation. *Journal of Comparative and Physiological Psychology*, **90**, 190–7.

Mittleman, G., Blaha, C. D. & Phillips, A. G. (1992). Pituitary–adrenal and dopaminergic modulation of schedule-induced polydipsia: behavioral and neurochemical evidence. *Behavioral Neuroscience*, **106**, 408–20.

Mogenson, G. J. (1977). *The neurobiology of behavior: an introduction.* Hillsdale, NJ: Lawrence Erlbaum Associates.

Mogenson, G. J., Jones, D. L. & Yim, C. Y. (1980). From motivation to action: functional interface between the limbic system and the motor system. *Progress in Neurobiology*, **14**, 69–97.

Møller, A. (1987). Social control of deception among status signalling house sparrows, Passer domesiticus. *Behavioral Ecology and Sociobiology*, **20**, 307–11.

Møller, A. P. (1988). Female choice selects for male sexual tail ornaments in the mongamous swallow. *Nature*, **332**, 640–2.

Møller, A. P. (1990). Parasites and sexual selection: current status of the Hamilton–Zuk hypothesis. *Journal of Evolutionary Biology*, **3**, 319–28.

Money, J. & Mathews, D. (1982). Prenatal exposure to virilizing progestins: an adult follow-up study of twelve women. *Archives of Sexual Behavior*, **11**, 73–83.

Montgomery, K. C. & Monkman, J. A. (1955). The relations between fear and exploratory behavior. *Journal of Comparative and Physiological Psychology*, **48**, 132–6.

Mook, D. G. (1996). *Motivation: the organization of action.* New York: W.W. Norton.

Moore, C. L. (1984). Maternal contributions to the development of masculine sexual behavior in laboratory rats. *Developmental Psychobiology*, **17**, 347–56.

Moore, C. L. (1985). Sex differences in urinary odors produced by young laboratory rats (Rattus norvegicus). *Journal of Comparative Psychology*, **99**, 336–41.

Moore, C. L. (1990). Comparative development of vertebrate sexual behavior: levels, cascades, and webs. In *Contemporary issues in comparative psychology*, ed. D. A. Dewsbury, pp. 278–99. Sunderland, MA: Sinauer.

Morgan, H. D., Fleming, A. S. & Stern, J. M. (1992). Somatosensory control of the onset and retention of maternal responsiveness in primiparous Sprague-Dawley rats. *Physiology and Behavior*, **51**, 549–55.

Morris, D. (1954). The reproductive behavior of the zebra finch (*Poephila guttata*) with special reference to pseudofemale behavior and displacement activities. *Behaviour*, **6**, 271–322.

Mowrer, O. H. (1960). *Learning theory and behavior.* New York: Wiley.

Moyer, K. E. (1977). A model of aggression with implications for research. (Proceedings.) *Psychopharmacology Bulletin*, **13**, 14–15.

Moyer, K. E. (1980). Biological substrates of aggression. *Progress in Brain Research*, **53**, 359–67.

Mrosovsky, N. (1990). *Rheostasis: the physiology of change.* New York: Oxford University Press.

Mykytowycz, R. (1985). Odour signals in the life of the wild rabbit. *Australian Science Magazine*, **1**, 41–4.

Mykytowycz, R. & Dudzinski, M. L. (1972). Aggressive and protective behaviour of adult rabbits *Oryctolagus cuniculus* towards juveniles. *Behaviour,* **43**, 97–120.

Nachman, M. (1962). Taste preference for sodium salts by adrenalectomized rats. *Journal of Comparative and Physiological Psychology,* **55**, 1124–9.

Nachman, M. (1963). Taste preferences for sodium salts by adrenalectomized rats. *American Journal of Physiology,* **205**, 219–21.

Nachman, M., Rauschenberger, J. & Ashe, J. H. (1977). Learned taste aversions over long delays in the rat: behavioral assessment of noxious drug effects. In *Food aversion learning,* ed. N. W. Milgram, L. Krames & T. M. Alloway, pp. 105–31. New York: Plenum Press.

New, R. S. & Benigni, L. (1987). Italian fathers and infants: cultural constraints on paternal behavior. In *The father's role: cross-cultural perspectives,* ed. M. E. Lamb, pp. 139–67. Hillsdale, NJ: Lawrence Erlbaum Associates.

Nicolaïdis, S. (1968). Responses of the hypothalamic osmosensitive units to saline and aqueous stimulations of the tongue. *Comptes Rendus de l'Académie des Sciences, Paris, Serie D,* **267**, 2352–5.

Nicolaïdis, S. & Rowland, N. (1975). Regulatory drinking in rats with permanent access to a bitter fluid source. *Physiology and Behavior,* **14**, 819–24.

Nisbett, R. E. (1968). Taste, deprivation, and weight determinants of eating behavior. *Journal of Personality and Social Psychology,* **10**, 107–16.

Nisbett, R. E. (1972). Hunger, obesity, and the ventromedial hypothalamus. *Psychological Review,* **79**, 433–53.

Noë, R., de Waal, F. B. & van Hooff, J. A. (1980). Types of dominance in a chimpanzee colony. *Folia Primatologica,* **34**, 90–110.

Novin, D., Robinson, K., Culbreth, L. A. & Tordoff, M. G. (1985). Is there a role for the liver in the control of food intake? *American Journal of Clinical Nutrition,* **42**, 1050–62.

Numan, M. (1996). A lesion and neuroanatomical tract-tracing analysis of the role of the bed nucleus of the stria terminalis in retrieval behavior and other aspects of maternal responsiveness in rats. *Developmental Psychobiology,* **29**, 23–51.

Numan, M., Rosenblatt, J. S. & Komisaruk, B. R. (1977). Medial preoptic area and onset of maternal behavior in the rat. *Journal of Comparative and Physiological Psychology,* **91**, 146–64.

O'Keefe, J. & Nadel, L. (1978). *The hippocampus as a cognitive map.* Oxford: Oxford University Press.

O'Malley, B. W. & Schrader, W. T. (1976). The receptors of steroid hormones. *Scientific American,* **234**, 32–43.

Oatley, K. (1971). Dissociation of the circadian drinking pattern from eating. *Nature,* **229**, 494–6.

Oatley, K. & Toates, F. M. (1973). Osmotic inhibition of eating as a subtractive process. *Journal of Comparative and Physiological Psychology,* **82**, 268–77.

Öhrwall, H. (1901). Die Modalitats- und Qualitatsbegriffe in der Sinnesphyiologie and dern Bedeutung. *Skandanavian Archiv für Physiologie,* **11**, 245–72.

Olds, J. & Olds, M. E. (1965). Drives, rewards, and the brain. In *New directions in psychology,* ed. F. Barron, W. C. Dement, W. Edwards & H. Lindmann, pp. 329–410. New York: Holt, Rinehart and Winston.

Oltmans, G. A. & Harvey, J. A. (1972). LH syndrome and brain catecholamine levels after lesions of the nigrostriatal bundle. *Physiology and Behavior,* **8**, 69–78.

Oomura, Y., Ono, T., Ooyama, H. & Wayner, M. J. (1969). Glucose and osmosensitive neurones of the rat hypothalamus. *Nature,* **222**, 282–4.

Orians, G. (1969). On the evolution of mating systems in birds and mammals. *American Naturalist,* **103**, 589–603.

Orpen, B. G., Furman, N., Wong, P. Y. & Fleming, A. S. (1987). Hormonal influences on the duration of postpartum maternal responsiveness in the rat. *Physiology and Behavior,* **40**, 307–15.

Packer, C. (1977). Reciprocal altruism in *Papio anubis. Nature,* **265**, 441–3.

Parsons, P. J., Fagan, T. & Spear, N. E. (1973). Short-term retention of habituation in the rat: a developmental study from infancy to old age. *Journal of Comparative and Physiological Psychology,* **84**, 545–53.

Passingham, R. E. (1982). *The human primate.* San Francisco: W.H. Freeman.

Paulus, R. A., Eng, R. & Schulkin, J. (1984). Preoperative latent place learning preserves salt appetite following damage to the central gustatory system. *Behavioral Neuroscience,* **98**, 146–51.

Peck, J. W. & Blass, E. M. (1975). Localization of thirst and antidiuretic osmoreceptors by intracranial injections in rats. *American Journal of Physiology,* **228**, 1501–9.

Pedersen, C. A. & Prange, A. J., Jr. (1979). Induction of maternal behavior in virgin rats after intracerebroventricular administration of oxytocin. *Proceedings of the National Academy of Sciences U S A,* **76**, 6661–5.

Pelchat, M. L. & Pliner, P. (1995). 'Try it. You'll like it': effects of information on willingness to try novel foods. *Appetite,* **24**, 153–65.

Perusse, D. (1993). Cultural and reproductive success in industrial societies: testing the relationship at the proximate and ultimate levels. *Behavioral and Brain Sciences,* **16**, 267–322.

Petri, H. L. (1996). *Motivation: theory, research, and applications.* Pacific Grove, CA: Brooks/Cole.

Pfaff, D. W. (1969). Histological differences between ventromedial hypothalamic neurones of well fed and underfed rats. *Nature,* **223**, 77–8.

Pfaff, D. W. (1982). *The physiological mechanisms of motivation.* New York: Springer-Verlag.

Pfeifer, W. D. & Davis, L. C. (1974). Effect of handling in infancy on responsiveness of adrenal tyrosine hydroxylase in maturity. *Behavioral Biology,* **10**, 239–45.

Phillips, A. G., Cox, V. C., Kakolewski, J. W. & Valenstein, E. S. (1969). Object-carrying by rats: an approach to the behavior produced by brain stimulation. *Science,* **166**, 903–5.

Phillips, A. G., Pfaus, J. & Blaha, C. (1991). Dopamine and motivated behavior: insights provided by in vivo analyses. In *The mesolimbic dopamine system: from motivation to action,* ed. P. Wilner & J. Scheel-Kruger, pp. 199–224. Chichester: Wiley.

Phillips, M. I. & Felix, D. (1976). Specific angiotensin II receptive neurons in the cat subfornical organ. *Brain Research,* **109**, 531–40.

Picco, G. (1999). *Man without a gun: one diplomat's secret struggle to free the hostages, fight terrorism, and a war.* New York: Times Books.

Pliner, P. (1982). The effects of mere exposure on liking for edible substances. *Appetite,* **3**, 283–90.

Pliner, P. & Hobden, K. (1992). Development of a scale to measure the trait of food neophobia in humans. *Appetite,* **19**, 105–20.

Pliner, P. & Pelchat, M. L. (1991). Neophobia in humans and the special status of foods of animal origin. *Appetite,* **16**, 205–18.

Poindron, P. (1974). Study of the mother–offspring relation in sheep (*Ovis aries*) at the time of suckling. *Comptes Rendus de l'Académie des Sciences, Paris, Serie D,* **278**, 2691–4.

Poindron, P., Le Neindre, P., Raksanyi, I., Trillat, G. & Orgeur, P. (1980). Importance of the characteristics of the young in the manifestation and establishment of maternal behaviour in sheep. *Reproduction Nutrition Development,* **20**, 817–26.

Porter, R. H., Cavallaro, S. A. & Moore, J. D. (1980). Developmental parameters of mother–offspring interactions in *Acomys cahinirus. Zeitschrift für Tierpsychologie,* **53**, 153–70.

Porter, R. H. & Doane, H. M. (1978). Studies of maternal behaviour in spiny mice (*Acomys cahirinus*). *Zeitschrift für Tierpsychologie,* **47**, 225–35.

Porter, R. H. & Lévy, F. (1995). Olfactory mediation of mother–infant interactions in selected mammalian species. In *Biological perspectives on motivated activities,* ed. R. Wong, pp. 77–110. Norwood, NJ: Ablex Publishing.

Powell, D. A., Schneiderman, N., Elster, A. J. & Jacobson, A. (1971). Differential classical conditioning in rabbits (*Oryctolagus cuniculus*) to tones and changes in illumination. *Journal of Comparative and Physiological Psychology,* **76**, 267–74.

Powers, W. T. (1978). Quantitative analysis of purposive systems: some spadework at the foundation of scientific psychology. *Psychological Review.* **85**, 417–35.

Powley, T. L. (1977). The ventromedial hypothalamic syndrome, satiety, and a cephalic phase hypothesis. *Psychological Review,* **84**, 89–126.

Powley, T. L. & Keesey, R. E. (1970). Relationship of body weight to the lateral hypothalamic feeding syndrome. *Journal of Comparative and Physiological Psychology,* **70**, 25–36.

Presser, H. B. (1978). Age at menarche, socio-sexual behavior, and fertility. *Social Biology,* **25**, 94–101.

Pruett-Jones, S. G., Pruett-Jones, M. A. & Jones, H. I. (1990). Parasites and sexual selection in birds of paradise. *American Zoologist,* **30**, 287–98.

Pryce, C. R. (1992). A comparative systems model of the regulation of maternal motivation in mammals. *Animal Behaviour,* **43**, 417–41.

Pryce, C. R. (1995). Determinants of motherhood in human and nonhuman primates: a biosocial model. In *Motherhood in human and nonhuman primates: biosocial determinants.,* ed. C. R. Pryce, R. D. Martin & D. Skuse, pp. 1–15. Basel: S. Karger.

Pryce, C. R., Abbott, D. H., Hodges, J. K. & Martin, R. D. (1988). Maternal behavior is related to prepartum urinary estradiol levels in red-bellied tamarin monkeys. *Physiology and Behavior,* **44**, 717–26.

Purtillo, D. T. & Sullivan, J. L. (1979). Immunological bases for superior survival of females. *American Journal of Diseases of Children,* **133**, 1251–3.

Raible, L. H. (1995). The biopsychology of feeding behavior in rodents: an attempt at integration. In *Biological perspectives on motivated activities,* ed. R. Wong, pp. 165–228. Norwood, NJ: Ablex Publishing.

Ramsay, D. J. & Reid, I. A. (1975). Some central mechanisms of thirst in the dog. *Journal of Physiology (London),* **253**, 517–25.

Ramsay, D. S., Seeley, R. J., Bolles, R. C. & Woods, S. C. (1996). Ingestive homeostasis: the primacy of learning. In *Why we eat what we eat,* ed. E. Capaldi & T. Powley, pp. 1–9. Washington, DC: American Psychological Association.

Rand, M. N. & Breedlove, S. M. (1987). Ontogeny of functional innervation of bulbocavernosus muscles in male and female rats. *Brain Research,* **430**, 150–2.

Raphelson, A. C., Isaacson, R. L. & Douglas, R. J. (1965). The effect of distracting stimuli on the runway performance of limbic damaged rats. *Psychonomic Science,* **3**, 483–4.

Raudenbush, B. & Frank, R. A. (1999). Assessing food neophobia: the role of stimulus familiarity. *Appetite,* **32**, 261–71.

Read, A. F. & Harvey, P. H. (1989). Reassessment of comparative evidence for Hamilton and Zuk theory on the evolution of secondary sexual characters. *Nature,* **339**, 618–20.

Reiman, E. M., Raichle, M. E., Butler, F. K., Herscovitch, P. & Robins, E. (1984). A focal brain abnormality in panic disorder, a severe form of anxiety. *Nature,* **310**, 683–5.

Reis, D. J., Doba, N. & Nathan, M. A. (1973). Predatory attack, grooming, and consummatory behaviors evoked by electrical stimulation of cat cerebellar nuclei. *Science*, **182**, 845–7.

Reite, M. & Boccia, M. L. (1994). Physiological aspects of adult attachment. In *Attachment in adults: clinical and developmental perspectives*, ed. M. B. Sperling & W. H. Berman, pp. 98–127. New York: Guilford Press.

Richardson, D. B. & Mogenson, G. J. (1981). Water intake elicited by injections of angiotensin II into preoptic area of rats. *American Journal of Physiology*, **240**, R70–4.

Richman, C. L. (1989). SAB, reward, and learning. In *Spontaneous alternation behavior*, ed. W. N. Dember & C. L. Richman, pp. 59–71. New York: Springer-Verlag.

Richter, C. P. (1936). Increased salt appetite in adrenalectomized rats. *American Journal of Physiology*, **115**, 155–61.

Richter, C. P. (1943). Total self-regulatory functions in animals and human beings. *The Harvey Lectures*, **38**, 63–103.

Richter, C. P. (1956). Salt appetite of animals: its dependence on instinct and metabolism. In *L'Instinct dans le comportement des animaux et de l'homme*, ed. M. Autorri, pp. 577–629. Paris: Masson.

Roberts, W. W. (1958). Rapid escape learning without avoidance learning motivated by hypothalamic stimulation in cats. *Journal of Comparative and Physiological Psychology*, **51**, 391–9.

Roberts, W. W. & Kiess, H. O. (1964). Motivational properties of hypothalamic aggression in cats. *Journal of Comparative and Physiological Psychology*, **61**, 187–93.

Robertson, G. L. (1976). The regulation of vasopressin function in health and disease. *Recent Progress in Hormone Research*, **33**, 333–85.

Robinson, J. E. & Short, R. V. (1977). Changes in breast sensitivity at puberty, during the menstrual cycle, and at parturition. *British Medical Journal*, **1**, 1188–91.

Robson, K. M. & Kumar, R. (1980). Delayed onset of maternal affection after childbirth. *British Journal of Psychiatry*, **136**, 347–53.

Rodgers, W. & Rozin, P. (1967). Novel food preferences in thiamine-deficient rats. *Journal of Comparative and Physiological Psychology*, **61**, 1–4.

Rodgers, W. L. (1967). Specificity of specific hungers. *Journal of Comparative and Physiological Psychology*, **64**, 49–58.

Rodin, J. (1981). Current status of the internal-external hypothesis for obesity: what went wrong? *American Psychologist*, **36**, 361–72.

Rodin, J., Moskowitz, H. R. & Bray, G. A. (1976). Relationship between obesity, weight loss, and taste responsiveness. *Physiology and Behavior*, **17**, 591–7.

Rolls, B. J. & Rolls, E. T. (1982). *Thirst*. Cambridge: Cambridge University Press.

Rolls, B. J., Wood, R. J., Rolls, E. T., Lind, H., Lind, W. & Ledingham, J. G. (1980). Thirst following water deprivation in humans. *American Journal of Physiology*, **239**, R476–82.

Rolls, B. J., Woods, S. C. & Rolls, E. T. (1980). Thirst: the initiation, maintenance, and termination of drinking. In *Progress in psychobiology and physiological psychology*, ed. J. M. Sprague & A. N. Epstein, pp. 263–321. New York: Academic Press.

Rolls, E. T. (1981a). The neurophysiology of feeding. *Proceedings of the Nutrition Society*, **40**, 361–2.

Rolls, E. T. (1981b). Central nervous mechanisms related to feeding and appetite. *British Medical Bulletin*, **37**, 131–4.

Rolls, E. T. (1984). The neurophysiology of feeding. *International Journal of Obesity and Related Dysfunctions*, **8**, 139–50.

Rolls, E. T. (1994). Neural processing related to feeding in primates. In *Appetite: neural and behavioural bases*, ed. C. R. Legg & D. A. Booth, pp. 11–53. Oxford: Oxford University Press.

Rolls, E. T. (1999). *The brain and emotion*. Oxford: Oxford University Press.

Rolls, E. T., Burton, M. J. & Mora, F. (1980). Neurophysiological analysis of brain-stimulation reward in the monkey. *Brain Research*, **194**, 339–57.

Ronmeyer, A. & Poindron, P. (1992). Early maternal discrimination of alien kids by postparturient goats. *Behaviour Processes*, **26**, 103–12.

Ronmeyer, A., Porter, R. H., Lévy, F., Nowak, R. Orgeur, R. & Poindron, P. (1993). Maternal labelling is not necessary for the establishment of discrimination between kids by recently parturient goats. *Animal Behaviour*, **46**, 706–12.

Rosenblatt, J. S. (1965). The basis of synchrony in the behavioural interaction between the mother and her offspring in the laboratory rat. In *Determinants of infant behaviour*, ed. B. M. Foss, pp. 3–41. London: Methuen.

Rosenblatt, J. S. (1967). Nonhormonal basis of maternal behavior in the rat. *Science*, **156**, 1512–4.

Rosenblatt, J. S. (1990). Landmarks in the physiological study of maternal behavior with special reference to the rat. In *Mammalian parenting: biochemical, neurobiological, and behavioral determinants*, ed. N. A. Krasnegor & R. S. Bridges, pp. 40–60. New York: Oxford University Press.

Rosenblatt, J. S. (1991). A psychobiological approach to maternal behaviour among the primates. In *The development and integration of behaviour: essays in honour of Robert Hinde*, ed. P. Bateson, pp. 191–222. Cambridge: Cambridge University Press.

Rosenblatt, J. S. & Lehrman, D. S. (1963). Maternal behavior of the laboratory rat. In *Maternal behavior in mammals*, ed. H. L. Rheingold, pp. 8–57. New York: Wiley.

Rosenblatt, J. S., Mayer, A. D. & Giordano, A. L. (1988). Hormonal basis during pregnancy for the onset of maternal behavior in the rat. *Psychoneuroendocrinology*, **13**, 29–46.

Rosenblatt, J. S. & Siegel, H. I. (1981). Factors governing the onset and maintenance of maternal behavior among primiparous mammals. In *Parental care in mammals*, ed. D. J. Gubernick & P. H. Klopfer, pp. 13–76. New York: Plenum Press.

Rosenblatt, J. S., Siegel, H. I. & Mayer, A. D. (1979). Progress in the study of maternal behavior in the rat: hormonal, nonhormonal, sensory, and developmental aspects. In *Advances in the study of behavior*, ed. C. G. Beer & M. C. Busnel, pp. 225–311. New York: Academic Press.

Rosenstein, D. & Oster, H. (1988). Differential facial responses to four basic tastes in newborns. *Child Development*, **59**, 1555–68.

Rosenthal, G. A., Janzen, D. H. & Applebaum, S. W. (1979). *Herbivores, their interaction with secondary plant metabolites*. New York: Academic Press.

Rothchild, I. (1962). The corpus luteum–pituitary relationship. The association between the cause of luteotrophin secretion and the cause of follicular quiesence during lactation: the basis for a tentative theory of the corpus luteum–pituitary relationship in the rat. *Endocrinology*, **67**, 9–41.

Routtenberg, A. (1968). The two-arousal hypothesis: reticular formation and limbic system. *Psychological Review*, **75**, 51–80.

Routtenberg, A. & Kuznesof, A. W. (1967). Self-starvation of rats living in activity wheels on a restricted feeding schedule. *Journal of Comparative and Physiological Psychology*, **64**, 414–21.

Rowland, N. E., Li, B. H. & Morien, A. (1996). Brain mechanisms and the physiology of feeding. In *Why we eat what we eat*, ed. J. Capaldi, pp. 173–204. Washington, DC: American Psychological Association.

Rowland, N. E. & Nicolaïdis, S. (1976). Metering of fluid intake and determinants of *ad libitum* drinking in rats. *American Journal of Physiology*, **231**, 1–8.

Rozin, P. (1969). Adaptive food sampling patterns in vitamin deficient rats. *Journal of Comparative and Physiological Psychology*, **69**, 126–32.

Rozin, P. (1976). The selection of food by rats, humans, and other animals. In *Advances in the study of behavior*, ed. J. S. Rosenblatt, R. A. Hinde, E. Shaw & C. Beer, pp. 21–76. New York: Academic Press.

Rozin, P. (1977). *The use of characteristic flavorings in human culinary practice*. Boulder, CO: Westview Press.

Rozin, P. (1982). Human food selection: the interaction of biology, culture, and individual experience. In *The psychobiology of human food selection*, ed. L. M. Barker, pp. 225–54. Westport, CN: AVI Publishing.

Rozin, P. (1996). Sociocultural influences on food selection. In *Why we eat what we eat*, ed. J. Capaldi, pp. 233–63. Washington, DC: American Psychological Association.

Rozin, P. & Kalat, J. W. (1971). Specific hungers and poison avoidance as adaptive specializations of learning. *Psychological Review*, **78**, 459–86.

Rozin, P. & Schulkin, J. (1990). Food selection. In *Neurobiology of food and fluid intake. Handbook of behavioral neurobiology*, ed. M. S. Edward, pp. 297–328. New York: Plenum Press.

Ruger, J. & Schulkin, J. (1980). Preoperative sodium appetite experience and hypothalamic lesions in rats. *Journal of Comparative and Physiological Psychology*, **94**, 914–20.

Russell, P. A. (1973). Relationships between exploratory behaviour and fear: a review. *British Journal of Psychology*, **64**, 417–33.

Russell, P. A. (1983). Pychological studies of exploration in animals. In *Explorations in animals and men*, ed. J. Archer & L. Birke, pp. 22–54. Wokingham: Van Nostrand.

Russell, P. A. & Williams, D. I. (1973). Effects of repeated testing on rats' locomotor activity in the open-field. *Animal Behaviour*, **21**, 109–11.

Russell, P. J., Abdelaal, A. E. & Mogenson, G. J. (1975). Graded levels of hemorrhage, thirst and angiotensin II in the rat. *Physiology and Behavior*, **15**, 117–19.

Rzoska, J. (1953). Bait shyness, a study in rat behaviour. *British Journal of Animal Behaviour*, **1**, 129–135.

Sackett, G. P. (1965). Effects of rearing conditions upon the behavior of rhesus monkeys (*Macaca mulatta*). *Child Development*, **36**, 855–68.

Salamone, J. D., Steinpreis, R. E., McCullough, L. D., Smith, P., Grebel, D. & Mahan, K. (1991). Haloperidol and nucleus accumbens dopamine depletion suppress lever pressing for food but increase free food consumption in a novel food choice procedure. *Psychopharmacology*, **104**, 515–21.

Salzen, E. A. (1962). Imprinting and fear. *Symposium of the Zoological Society of London*, **8**, 184–9.

Salzen, E. A. (1970). Imprinting and environmental learning. In *Development and evolution of behavior*, ed. L. R. Aronson, E. Tobach & J. S. Rosenblatt, pp. 158–78. San Francisco: W. H. Freeman.

Sarkar, S. (1984). Motility, expression of surface antigen, and X and Y human sperm separation in in vitro fertilization medium. *Fertility and Sterility*, **42**, 899–905.

Schachter, S. (1968). Obesity and eating. *Science*, **161**, 751–6.

Schachter, S. (1971). Some extraordinary facts about obese humans and rats. *American Psychologist*, **26**, 129–44.

Schachter, S. & Gross, L. P. (1968). Manipulated time and eating behavior. *Journal of Personality and Social Psychology*, **10**, 98–106.

Schachter, S. & Rodin, J. (1974). *Obese humans and rats*. Potomac, MD: Lawrence Erlbaum Associates.

Schallert, T. & Wishaw, I. Q. (1978). Two types of sensorimotor impairment after lateral hypothalamic lesions: observatons in normal weight, dieted, and fattened rats. *Journal of Comparative and Physiological Psychology*, **92**, 720–41.

Schneirla, T. C. (1965). *Aspects of stimulation and organization in approach/withdrawal processes underlying vertebrate development*. New York: Academic Press.

Schulkin, J. (1982). Behavior of sodium-deficient rats: the search for a salty taste. *Journal of Comparative and Physiological Psychology*, **96**, 629–34.

Schulkin, J. (1986). The evolution and expression of salt appetite. In *The physiology of thirst and sodium appetite*, ed. G. deCaro, A. N. Epstein & M. Massi, pp. 491–6. New York: Plenum Press.

Schulkin, J. (1989). The effects of preoperative alimentary experiences on a regulatory neurobehavioral system. In *Preoperative events: their effects on behavior following brain damage. Comparative cognition and neuroscience*, ed. J. Schulkin, pp. 21–34. Hillsdale, NJ: Lawrence Erlbaum Associates.

Schulkin, J. (1999). *The neuroendocrine regulation of behavior*. Cambridge: Cambridge University Press.

Schulze, G. (1995). Motivation: homeostatic mechanisms may instigate and shape adaptive behaviors through the generation of hedonic states. In *Biological perspectives on motivated activities*, ed. R. Wong, pp. 265–88. Norwood, NJ: Ablex Publishing.

Schwalb, D. W., Imaizumi, N. & Nakazawa, J. (1987). The modern Japanese father: roles and problems in a changing society. In *The father's role: cross-cultural perspectives.*, ed. M. E. Lamb, pp. 247–68. Hillsdale, NJ: Lawrence Erlbaum Associates.

Sclafani, A. (1981). The role of hyperinsulinema and the vagus nerve in hypothalamic hyperphagia reexamined. *Diabetologia*, **20** (Suppl.), 402–10.

Scouten, C. W., Groteleuschen, L. K. & Beatty, W. W. (1975). Androgens and the organization of sex differences in active avoidance behavior in the rat. *Journal of Comparative and Physiological Psychology*, **88**, 264–70.

Seltzer, C. P. (1998). The use of investigatory responses as a measure of learning and memory. In *Human factors and web development*, ed. C. Forsythe & E. Grose, pp. 17–24. Mahwah, NJ: Lawrence Erlbaum Associates.

Seyfarth, R. M. & Cheney, D. L. (1984). Grooming, alliances and reciprocal altruism in vervet monkeys. *Nature*, **308**, 541–3.

Sheldon, A. B. (1969). Preference for familiar novel stimuli as a function of familiarity of the environment. *Journal of Comparative and Physiological Psychology*, **67**, 516–21.

Shephard, R. A. & Estall, L. B. (1984a). Anxiolytic actions of chlordiazepoxide determine its effects on hyponeophagia in rats. *Psychopharmacology*, **82**, 343–7.

Shephard, R. A. & Estall, L. B. (1984b). Effects of chlordiazepoxide and of valproate on hyponeophagia in rats. Evidence for a mutual antagonism between their anxiolytic properties. *Neuropharmacology*, **23**, 677–81.

Sherman, P. W. (1977). Nepotism and the evolution of alarm calls. *Science*, **197**, 1246–53.

Sherman, P. W. (1985). Alarm calls of Belding's ground squirrels to aerial predators: nepotism or self-preservation? *Behavioral Ecology and Sociobiology*, **17**, 313–23.

Shillito, E. E. (1963). Exploratory behaviour in the short-tailed vole, *Microtus agrestis*. *Behaviour*, **21**, 145–54.

Short, R. V. (1979). Sexual selection and its component parts: somatic and genital selection, as illustrated by man and the great apes. In *Advances in the study of behavior*, ed. J. Rosenblatt, R. A. Hinde, C. Beer & M.-C. Busnel, pp. 131–58. New York: Academic Press.

Short, R. V. (1994). Why sex? In *The differences between the sexes*, ed. R. V. Short & E. Balaban, pp. 3–22. Cambridge: Cambridge University Press.

Simoons, F. J. (1970). Primary adult lactose intolerance and the milking habit: A problem in biologic and cultural interrelations. II. A culture historical hypothesis. *American Journal of Digestive Disorders*, **15**, 695–710.

Simoons, F. J. (1978). The geographic hypothesis and lactose malabsorption. A weighing of the evidence. *American Journal of Digestive Disorders*, **23**, 963–80.

Simoons, F. J. (1982). A geographic approach to senile cataracts: possible links with milk consumption, lactase activity, and galactose metabolism. *Digestive Diseases and Sciences*, **27**, 257–64.

Simpson, J. B., Epstein, A. N. & Camardo, J. S., Jr. (1978). Localization of receptors for the dipsogenic action of angiotensin II in the subfornical organ of rat. *Journal of Comparative and Physiological Psychology*, **92**, 581–601.

Simpson, J. B. & Routtenberg, A. (1973). Subfornical organ: site of drinking elicitation by angiotensin II. *Science*, **181**, 1772–5.

Singh, S. D. (1969). Urban monkeys. *Scientific American*, **221**, 108–15.

Smith, H. G. & Montgomerie, R. (1991). Sexual selection and tail ornaments of North American barn swallows. *Behavioural Ecology and Sociobiology*, **8**, 171–8.

Smith, R. L. (1984). *Sperm competition and the evolution of animal mating systems*. Orlando: Academic Press.

Smotherman, W. P., Wiener, S. G., Mendoza, S. P. & Levine, S. (1977). Maternal pituitary–adrenal responsiveness as a function of differential treatment of rat pups. *Developmental Psychobiology*, **10**, 113–22.

Snowdon, C. T. (1990). *Mechanisms maintaining monogamy in monkeys*. Sunderland, MA: Sinauer Associates.

Snowdon, C. T. & Epstein, A. N. (1970). Oral and intragastric feeding in vagotomized rats. *Journal of Comparative and Physiological Psychology*, **71**, 59–67.

Snowdon, C. T. & Suomi, S. J. (1982). Paternal behavior in infants. In *Child nurturance*, ed. H. E. Fitzgerald, J. A. Mullins & P. Gage, pp. 63–108. New York: Plenum Press.

Snowdon, C. T. & Wampler, R. S. (1979). Weight and regulatory deficits in vagotomized rats: a reexamination. *Behavioral and Neural Biology*, **26**, 342–53.

Solomon, R. & Wynne, L. C. (1954). Traumatic avoidance learning: the principles of anxiety conservation and partial irreversibility. *Psychological Review*, **61**, 353–85.

Spear, N. E. (1978). *The processing of memories: forgetting and retention*. Hillsdale, NJ: Lawrence Erlbaum Associates.

Spear, N. E. & Hill, W. F. (1962). Methodological note: excessive weight loss in activity wheels. *Psychological Reports*, **11**, 437–8.

Spear, N. E. & Miller, J. S. (1989). Ontogeny of spontaneous alternation behavior. In *Spontaneous alternation behavior*, ed. W. N. Dember & C. L. Richman, pp. 131–44. New York: Springer-Verlag.

Spiegel, T. A., Shrager, E. E. & Stellar, E. (1989). Responses of lean and obese subjects to preloads, deprivation, and palatability. *Appetite*, **13**, 45–69.

Sroufe, L. A. (1977). Wariness of strangers and the study of infant development. *Child Development*, **48**, 1184–99.

Sroufe, L. A. & Waters, E. (1977). Attachment as an organized construct. *Child Development*, **48**, 55–71.

Staddon, J. E. R. (1983). *Adaptive behavior and learning*. Cambridge: Cambridge University Press.

Staddon, J. E. R. & Simmelhag, V. L. (1971). The 'superstition' experiment: a rexamination of its implications fro the principle of adaptive behavior. *Psychological Review*, **78**, 3–43.

Stanley, B. G., Daniel, D. R., Chin, A. S. & Leibowitz, S. F. (1985). Paraventricular nucleus injections of peptide YY and neuropeptide Y preferentially enhance carbohydrate ingestion. *Peptides*, **6**, 1205–11.

Stanley, B. G., Magdalin, W., Seirafi, A., Thomas, W. J. & Leibowitz, S. F. (1993). The perifornical area: the major focus of (a) patchily distributed hypothalamic neuropeptide Y-sensitive feeding system(s). *Brain Research*, **604**, 304–17.

Stanley, B. G., Schwartz, D. H., Hernandez, L., Hoebel, B. G. & Leibowitz, S. F. (1989). Patterns of extracellular norepinephrine in the paraventricular hypothalamus: relationship to circadian rhythm and deprivation-induced eating behavior. *Life Sciences*, **45**, 275–82.

Steiner, J. E. (1979). Human facial expressions in response to taste and smell stimulation. *Advances in Child Development and Behavior*, **13**, 257–95.

Stellar, E. (1954). The physiology of motivation. *Psychological Review*, **61**, 5–22.

Stellar, E. (1990). Brain and behavior. In *Neurobiology of food and fluid intake. Handbook of behavioral neurobiology*, vol. 10, ed. M. S. Edward, pp. 3–22. New York: Plenum Press.

Stellar, J. R., Brooks, F. H. & Mills, L. E. (1979). Approach and withdrawal analysis of the effects of hypothalamic stimulation and lesions in rats. *Journal of Comparative and Physiological Psychology*, **93**, 446–66.

Stellar, J. R. & Stellar, E. (1985). *The neurobiology of motivation and reward*. New York: Springer-Verlag.

Stephan, F. K., Valenstein, E. S. & Zucker, I. (1971). Copulation and eating during electrical stimulation of the rat hypothalamus. *Physiology and Behavior*, **7**, 587–93.

Stern, K. & McClintock, M. K. (1998). Regulation of ovulation by human pheromones. *Nature*, **392**, 177–9.

Stevens, R. & Cowey, A. (1973). Effects of dorsal and ventral hippocampal lesions on spontaneous alternation, learned alternation and probability learning in rats. *Brain Research*, **52**, 203–24.

Stevenson-Hinde, J. (1991). Temperament and attachment: an eclectic approach. In *The development and integration of behaviour: essays in honour of Robert Hinde*, ed. P. Bateson, pp. 315–29. Cambridge: Cambridge University Press.

Stewart, J. (1988). Current themes, theoretical issues, and preoccupations in the study of sexual differentiation and gender-related behaviors. Special Issue: sexual differentiation and gender-related behaviors. *Psychobiology*, **16**, 315–20.

Stewart, J. & Cygan, D. (1980). Ovarian hormones act early in development to feminize adult open-field behavior in the rat. *Hormones and Behavior*, **14**, 20–32.

Stewart, J. & Kolb, B. (1988). The effects of neonatal gonadectomy and prenatal stress on cortical thickness and asymmetry in rats. *Behavioral and Neural Biology*, **49**, 344–60.

Stolk, J. M., Conner, R. L., Levine, S. & Barchas, J. D. (1974). Brain norepinephrine metabolism and shock-induced fighting behavior in rats: differential effects of shock and fighting on the neurochemical response to a common footshock stimulus. *Journal of Pharmacology and Experimental Therapeutics*, **190**, 193–209.

Stricker, E. M. (1976). Drinking by rats after lateral hypothalamic lesions. *Journal of Comparative and Physiological Psychology*, 90, 127–43.

Stricker, E. M. (1978). Hyperphagia. *New England Journal of Medicine*, **298**, 1010–13.

Stricker, E. M. (1990). Homeostatic origins of ingestive behavior. In *Neurobiology of food and fluid intake. Handbook of behavioral neurobiology*, ed. E. M. Stricker, pp.

45–60. New York: Plenum Press.

Stricker, E. M. & Sterritt, G. M. (1967). Osmoregulation in the newly hatched domestic chick. *Physiology and Behavior,* **2**, 117–19.

Stricker, E. M. & Verbalis, J. G. (1990). Sodium appetite. In *Neurobiology of food and fluid intake. Handbook of behavioral neurobiology,* ed. E. M. Stricker, pp. 387–419. New York: Plenum Press.

Stricker, E. M. & Verbalis, J. G. (1999). Water intake and body fluids. In *Fundamental neuroscience,* ed. M. J. Zigmond *et al.*, pp. 1091–109. San Diego: Academic Press.

Stricker, E. M. & Zigmond, M. J. (1975). Brain catecholamines and the lateral hypothalamic syndrome. In *Hunger: basic mechanisms and clinical implications,* ed. D. Novin, W. Wyrwicka & G. A. Bray, pp. 19–32. New York: Raven Press.

Stunkard, A. J. & Fox, S. (1971). The relationship of gastric motility and hunger: a summary of the evidence. *Psychosomatic Medicine,* **33**, 123–34.

Suarez, S. D. & Gallup, G. G. (1985). Open-field behaviour in chickens: a replication revisited. *Behavioural Processes,* **10**, 333–40.

Sumner, A. T. & Robinson, J. A. (1976). A difference in dry mass between the heads of X- and Y-bearing human spermatozoa. *Journal of Reproduction and Fertility,* **48**, 9–15.

Surbey, M. K. (1987). Anorexia nervosa, amenorrhea, and adaptation. *Ethology and Sociobiology,* **8**, 47–61.

Swanson, L. W., Kucharczyk, J. & Mogenson, G. J. (1978). Autoradiographic evidence for pathways from the medial preoptic area to the midbrain involved in the drinking response to angiotensin II. *Journal of Comparative Neurology,* **178**, 645–59.

Symons, D. (1979). *The evolution of human sexuality.* New York: Oxford University Press.

Tardif, S. D., Richter, C. B. & Carson, R. L. (1984). Effects of sibling-rearing experience on future reproductive success in two species of Callitrichidae. *American Journal of Primatology,* **6**, 377–80.

Teitelbaum, P. (1955). Sensory control of hypothalamic hyperphagia. *Journal of Comparative and Physiological Psychology,* **50**, 486–90.

Teitelbaum, P. & Epstein, A. N. (1962). The lateral hypothalamic syndrome. *Psychological Review,* **69**, 74–90.

Tempel, D. L., Leibowitz, K. J. & Leibowitz, S. F. (1988). Effects of PVN galanin on macronutrient selection. *Peptides,* **9**, 309–14.

Terkel, J. & Rosenblatt, J. S. (1968). Maternal behavior induced by maternal blood plasma injected into virgin rats. *Journal of Comparative and Physiological Psychology,* **65**, 479–82.

Terkel, J. & Rosenblatt, J. S. (1972). Humoral factors underlying maternal behavior at parturition: cross transfusion between freely moving rats. *Journal of Comparative and Physiological Psychology,* **80**, 365–71.

Thierry, B. & Anderson, J. R. (1986). Adoption in anthropoid primates. *International Journal of Primatology,* **7**, 191–216.

Thompson, C. W., Hillgarth, N., Leu, M. & McClure, H. E. (1997). High parasite load in house finches (*Carpadocacus mexicanus*) is highly correlated with reduced expression of a sexually selected trait. *American Naturalist,* **149**, 270–94.

Thornton, J. & Goy, R. W. (1986). Female-typical sexual behavior of rhesus and defeminization by androgens given prenatally. *Hormones and Behavior,* **20**, 129–47.

Thrasher, T. N., Keil, L. C. & Ramsay, D. J. (1982). Lesions of the organum vasculosum of the lamina terminalis (OVLT) attenuate osmotically-induced drinking and vasopressin secretion in the dog. *Endocrinology,* **110**, 1837–9.

Timmermans, P. J., Roder, E. L. & Hunting, P. (1986). The effect of absence of the mother on the acquisition of phobic behaviour in cynomolgus monkeys (*Macaca fascicularis*). *Behaviour Research and Therapy*, **24**, 67–72.

Timmermans, P. J. A., Vochteloo, J. D., Vossen, J. M. H. *et al.* (1994). Persistent neophobic behaviour in monkeys: a habit or a trait? *Behavioural Processes*, **31**, 177–96.

Tinbergen, N. (1951). *The study of instinct.* Oxford: Clarendon Press.

Tinbergen, N. (1954). Some neurophysiological problems raised by ethology. *British Journal of Animal Behaviour*, **2**, 115.

Tinbergen, N. (1968). On war and peace in animals and man. An ethologist's approach to the biology of aggression. *Science*, **160**, 1411–18.

Toates, F. M. (1980). *Animal behaviour: a systems approach.* Chichester: Wiley.

Toates, F. M. (1983). Exploration as a motivational and learning system. In *Explorations in animals and humans*, ed. J. Archer & L. Birke, pp. 55–71. Wokingham: Van Nostrand.

Toates, F. M. (1986). *Motivational systems.* Cambridge: Cambridge University Press.

Toates, F. M. (1988). Motivation and emotion from a biological perspective. In *Cognitive perspectives on emotion and motivation. NATO ASI series D. Behavioural and social sciences*, ed. V. Hamilton, G. H. Bower & N. H. Frijda, pp. 3–35. Dordrecht, Kluwer Academic Publishers.

Toates, F. M. & Booth, D. A. (1974). Control of food intake by energy supply. *Nature*, **251**, 710–11.

Tolman, E. C. (1948). Cognitive maps in rats and men. *Psychological Review*, **55**, 189–208.

Tooby, J. & Cosmides, L. (1990a). On the universality of human nature and the uniqueness of the individual: the role of genetics and adaptation. Special Issue. Biological foundations of personality: evolution, behavioral genetics, and psychophysiology. *Journal of Personality*, **58**, 17–67.

Tooby, J. & Cosmides, L. (1990b). The past explains the present: emotional adaptations and the structure of ancestral environments. *Ethology and Sociobiology*, **11** (Special Issue), 375–424.

Toran-Allerand, C. D. (1984). Gonadal hormones and brain development: implications for the genesis of sexual differentiation. *Annals of the New York Academy of Sciences*, **435**, 101–10.

Trivers, R. L. (1972). Parental investment and sexual selection. In *Sexual selection and the descent of man 1871–1971*, ed. B. Campbell, pp. 136–79. Chicago: Aldine.

Trivers, R. L. (1974). Parent–offspring conflict. *American Zoologist*, **14**, 249–64.

Tryon, R. C. (1931). Studies in individual differences in maze ability. II. The determination of individual differences by age, weight, sex, and pigmentation. *Journal of Comparative Psychology*, **12**, 1–22.

Turner, S. G., Sechzer, J. A. & Liebelt, R. A. (1967). Sensitivity to electric shock after ventromedial hypothalamic lesions. *Experimental Neurology*, **19**, 236–44.

Udry, J. R. & Cliquet, R. L. (1982). A cross-cultural examination of the relationship between ages at menarche, marriage, and first birth. *Demography*, **19**, 53–63.

Ungerstedt, U. (1971). Adipsia and aphagia after 6-hydroxydopamine induced degeneration of the nigro-striatal dopamine system. *Acta Physiologica Scandinavica* (Supplementum), **367**, 95–122.

Valenstein, E. S. (1973). *Brain stimulation and motivation: research and commentary.* Glenview, IL: Scott Foresman.

Valenstein, E. S., Cox, V. C. & Kakolewski, J. W. (1968). Modification of motivated behavior elicited by electrical stimulation of the hypothalamus. *Science*, **159**, 1119–21.

Valenstein, E. S., Cox, V. C. & Kakolewski, J. W. (1970). Reexamination of the role of the hypothalamus in motivation. *Psychological Review*, **77**, 16–31.

Valenstein, E. S., Kakolewski, J. W. & Cox, V. C. (1967). Sex differences in taste preference for glucose and saccharin solutions. *Science*, **156**, 942–3.

Valle, F. P. (1972). Free and forced exploration in rats as a function of between vs within Ss design. *Psychonomic Science*, **23**, 333–5.

Valle, F. P. (1995). A reexamination of the role of associative factors in the control of normal drinking in the rat. In *Biological perspectives on motivated activities*, ed. R. Wong, pp. 289–335. Norwood, NJ: Ablex Publishing.

van Aberleen, J. H. F. (1989). Genetic control of hippocampal and dynophinergic mechanisms regulating novelty-induced exploratory behavior in house mice. *Experientia*, **45**, 839–45.

Van den Berghe, P. L. (1975). *Man in society: a biosocial view*. New York: Elsevier.

Van Itallie, T. B. (1984). The enduring storage capacity for fat: implications for treatment of obesity. *Research Publication of the Association for Research in Nervous Mental Disease*, **62**, 109–19.

Verbalis, J., G,, Hoffman, G. E. & Sherman, T. G. (1995). Use of immediate early genes as markers of oxytocin and vasopressin neuronal factors. *Current Opinion in Endocrine Metabolism*, **2**, 157–68.

Verbalis, J. G. (1990). Clinical aspects of body fluid homeostasis in humans. In *Neurobiology of food and fluid intake. Handbook of behavioral neurobiology*, ed. E. M. Stricker, pp. 421–62. New York: Plenum Press.

Verney, E. G. (1947). The antidiuretic hormone and the factors which determine its release. *Proceedings of the Royal Society of London (B)*, **135**, 25–106.

Vilberg, T. R. & Beatty, W. W. (1975). Behavioral changes following VMH lesions in rats with controlled insulin levels. *Pharmacology, Biochemistry and Behavior*, **3**, 377–84.

Vincent, J. D., Arnauld, E. & Bioulac, B. (1972). Activity of osmosensitive single cells in the hypothalamus of the behaving monkey during drinking. *Brain Research*, **44**, 371–84.

Vining, D. R. (1986). Social and reproductive success: the central theoretical problem of human sociobiology. *Behavioral and Brain Sciences*, **9**, 167–87.

Vochteloo, J. D., Timmermans, P. J., Duijghuisen, J. A. & Vossen, J. M. (1991). Responses to novelty in phobic and non-phobic cynomolgus monkeys: the role of subject characteristics and object features. *Behaviour Research and Therapy*, **29**, 531–8.

Waal, F. B. M. de. (1982). *Chimpanzee politics: power and sex among apes*. New York: Harper & Row.

Wade, G. N. & Zucker, I. (1969). Hormonal and developmental influences on rat saccharin preferences. *Journal of Comparative and Physiological Psychology*, **69**, 291–300.

Wagner, A. R. (1963). Conditioned frustration as a learned drive. *Journal of Experimental Psychology*, **66**, 142–8.

Wallace, R. A. (1979). *Animal behavior, its development, ecology, and evolution*. Santa Monica, CA: Goodyear Publishing.

Walsh, B. T., Katz, J. L., Levin, J., Kream, J., Fukushima, D. K., Hellman, L. D., Weiner, H. & Zumoff, B. (1978). Adrenal activity in anorexia nervosa. *Psychosomatic Medicine*, **40**, 499–506.

Washburn, S. L. & Lancaster, C. S. (1968). The evolution of hunting. In *Man the hunter*, ed. I. DeVore, pp. 293–303. Chicago: Aldine.

Wasman, M. & Flynn, J. P. (1962). Directed attack elicited from hypothalamus. *Archives of Neurology*, **6**, 220–7.

Wasser, S. K. & Barash, D. P. (1983). Reproductive suppression among female mammals: implications for biomedicine and sexual selection theory. *Quarterly Review of Biology,* **58**, 513–38.

Watanabe, K., Hara, C. & Ogawa, N. (1992). Feeding conditions and estrous cycle of female rats under the activity-stress procedure from aspects of anorexia nervosa. *Physiology and Behavior,* **51**, 827–32.

Weinberg, J., Erskine, M. & Levine, S. (1980). Shock-induced fighting attenuates the effects of prior shock experience in rats. *Physiology and Behavior,* **25**, 9–16.

Weinberg, J., Smotherman, W. P. & Levine, S. (1978). Early handling effects on neophobia and conditioned taste aversion. *Physiology and Behavior,* **20**, 589–96.

Weingarten, H. P. & Powley, T. L. (1980). Ventromedial hypothalamic lesions elevate basal and cephalic phase gastric acid output. *American Journal of Physiology,* **239**, G221–9.

Weiner, B. (1980). *Human motivation.* New York: Holt Rinehart and Winston.

Weiner, B. (1985). *Human motivation.* 2nd edn. New York: Springer-Verlag.

Weingarten, H. P. & Powley, T. L. (1980). Ventromedial hypothalamic lesions elevate basal and cephalic phase gastric acid output. *American Journal of Physiology,* **237**, G221–9.

Weisinger, R. S. & Woods, S. C. (1971). Aldosterone-elicited sodium appetite. *Endocrinology,* **89**, 538–44.

Weisinger, R. S., Woods, S., C. & Skorupski, J. D. (1970). Sodium deficiency and latent learning. *Psychonomic Science,* **19**, 307–8.

Weiss, J. M., Pohorecky, L. A., Salman, S. & Gruenthal, M. (1976). Attenuation of gastric lesions by psychological aspects of aggression in rats. *Journal of Comparative and Physiological Psychology,* **90**, 252–9.

Wells, P. A., Lowe, G., Sheldon, M. H. & Williams, D. I. (1969). Effects of infantile stimulation and environmental familiarity on exploratory behaviour in the rat. *British Journal of Psychology,* **60**, 389–93.

Whimbey, A. E. & Denenberg, V. H. (1967a). Experimental programming of life histories: the factor structure underlying experimentally created individual differences. *Behaviour,* **29**, 296–314.

Whimbey, A. E. & Denenberg, V. H. (1967b). Two independent behavioral dimensions in open-field performance. *Journal of Comparative and Physiological Psychology,* **63**, 500–4.

Wilkins, L. & Richter, C. P. (1940). A great craving for salt by a child with cortico-adrenal deficiency. *Journal of the American Medical Association,* **114**, 866–8.

Wilkinson, G. S. (1984). Reciprocal food sharing in the vampire bat. *Nature,* **308**, 181–4.

Williams, D. I. (1972). Effects of electric shock on exploratory behavior in the rats. *Quarterly Journal of Experimental Psychology,* **24**, 544–6.

Williams, D. I. (1973). Infantile stimulation and exploratory behavior in the rat. *Developmental Psychobiology,* **6**, 393–7.

Williams, D. I. & Russell, P. A. (1972). Open-field behaviour in rats: effects of handling sex and repeated testing. *British Journal of Psychology,* **63**, 593–6.

Williams, G. C. (1966). *Adaptation and natural selection: a critique of some current evolutionary thought.* Princeton, NJ: Princeton University Press.

Williams, J. H. & Gray, J. A. (1996). Dependence of the proactive behavioral effects of theta-driving septal stimulation on stimulation. 1. Leverpress experiments. *Psychobiology,* **24**, 9–21.

Wilson, E. O. (1975). *Sociobiology: the new synthesis.* Cambridge, MA: Belknap Press of Harvard University Press.

Wilson, M. (1993). What is the adaptation: status striving, status itself or parental teaching biases? *Behavioral and Brain Sciences,* **16**, 311.

Wilson, M., Johnson, H. & Daly, M. (1995). Lethal and nonlethal violence against wives. Special Issue. Focus on the Violence Against Women Survey. *Canadian Journal of Criminology,* **37**, 331–61.

Winn, P., Tarbuck, A. & Dunnett, S. B. (1984). Ibotenic acid lesions of the lateral hypothalamus: comparison with the electrolytic lesion syndrome. *Neuroscience,* **12**, 225–40.

Wise, R. A. (1982). Neuroleptics and operant behavior: the anhedonia hypothesis. *Behavioural and Brain Sciences,* **5**, 39–87.

Wise, R. A. (1994). A brief history of the anhedonia hypothesis. In *Appetite: neural and behavioural bases,* ed. C. L. Legg & D. A. Booth. Oxford: Oxford University Press.

Wise, R. A. & Raptis, L. (1986). Effects of naloxone and pimozide on initiation and maintenance measures of free feeding. *Brain Research,* **368**, 62–8.

Wittenberger, J. F. (1981). *Animal social psychology.* Boston: Duxbury.

Wolf, G. (1965). Effect of deoxycorticosterone on sodium appetite of intact and adenalectomized rats. *American Journal of Physiology,* **208**, 1281–5.

Wolf, G. (1967). Hypothalamic regulation of sodium intake: relations to preoptic and tegmental function. *American Journal of Physiology,* **213**, 1433–8.

Wolf, G. (1968). Thalamic and tegmental mechanisms for sodium intake. *Physiology and Behavior,* **6**, 997–1002.

Wolf, G. (1969). Innate mechanism for regulation of sodium intake. In *Olfaction and taste,* ed. C. Pfaffman, pp. 548–53. New York: Rockefeller University Press.

Wolf, G. & Handal, P. J. (1966). Aldosterone-induced sodium appetite: dose-response and specificity. *Endocrinology,* **78**, 1120–4.

Wolf, G. & Schulkin, J. (1980). Brain lesions and sodium appetite: an approach to the neurological analysis of homeostatic behavior. In *Biological and behavioral aspects of salt intake,* ed. M. R. Kare, M. J. Fregly & R. A. Bernard, pp. 331–9. New York: Academic Press.

Wolf, G., Schulkin, J. & Fluharty, S. J. (1983). Recovery of salt appetite after lateral hypothalamic lesions: effects of preoperative salt drive and salt intake experiences. *Behavioral Neuroscience,* **97**, 506–11.

Wong, R. (1969). Alternation rate as a function of infantile handling and food deprivation. *Journal of Genetic Psychology,* **115**, 237–46.

Wong, R. (1976). *Motivation: a biobehavioral analysis of consummatory activities.* New York: MacMillan.

Wong, R. (1994). Response latency of gerbils and hamsters to nuts flavoured with bitter-tasting substances. *Quarterly Journal of Experimental Psychology,* **47B**, 173–86.

Wong, R. (1995). Flavor neophobia in selected rodent species. In *Biological perspectives on motivated activities.* ed. R. Wong, pp. 229–46. Norwood, NJ: Ablex Publishing.

Wong, R. & Bowles, L. (1976). Exploration of complex stimuli as facilitated by emotional reactivity and shock. *American Journal of Psychology,* **89**, 527–34.

Wong, R. & Jelliffe, D. (1972). Infantile handling and food versus sensory-activity incentive choice. *Behavioral Biology,* **7**, 815–22.

Wong, R. & McBride, C. B. (1993). Flavour neophobia in gerbils (*Meriones unguiculatus*) and hamsters (*Mesocricetus auratus*). *Quarterly Journal of Experimental Psychology,* **46B**, 129–43.

Woods, S. C. (1991). The eating paradox: how we tolerate food. *Psychological Review,* **98**, 488–505.

Woods, S. C., Decke, E. & Vasselli, J. R. (1974). Metabolic hormones and regulation of body weight. *Psychological Review,* **81**, 26–43.

Woods, S. C. & Kulkosky, P. J. (1976). Classically conditioned changes of blood glucose level. *Psychosomatic Medicine,* **38**, 201–19.

Woods, S. C., Rolls, B. & Ramsay, D. J. (1977). Drinking following intracarotid infusions of hypertonic saline in dogs. *American Journal of Physiology*, **232**, R88–92.

Woods, S. C. & Stricker, E. M. (1999). Food intake and metabolism. In *Fundamental neuroscience*, ed. M. J. Zigmond, pp. 1091–109. San Diego: Academic Press.

Woods, S. C., Taborsky, G. J. & Porte, D. (1986). CNS control of nutrient homeostasis. In *Handbook of physiology*. Section 1. *The nervous system*, ed. F. Bloom, pp. 365–411. Bethesda, MD: American Physiological Society.

Wyrwicka, W. (1981). *The development of food preferences: parental influences and the primary effect*. Springfield, IL: Thomas.

Yahr, P. (1988). Pars compacta of the sexually dimorphic area of the gerbil hypothalamus: postnatal ages at which development responds to testosterone. *Behavioral and Neural Biology*, **49**, 118–24.

Yahr, P. & Greene, S. B. (1992). Effects of unilateral hypothalamic manipulations on the sexual behaviors of rats. *Behavioral Neuroscience*, **106**, 698–709.

Yamamoto, J. (1998). Relationship between hippocampal theta-wave frequency and emotional behavior in rabbits produced with stress or psychotropic drugs. *Japanese Journal of Pharmacology*, **76**, 125–7.

Young, P. T. (1948). Appetite, palatability and feeding habit: a critical review. *Psychological Review*, **45**, 289–320.

Young, P. T. (1961). *Motivation and emotion: a survey of the determinants of human and animal activity*. New York: Wiley.

Young, W. C., Goy, R. W. & Phoenix, C. H. (1964). Hormones and sexual behavior. *Science*, **143**, 212–18.

Zagrodzka, J. & Fonberg, E. (1978). Predatory versus alimentary behavior after amygdala lesions in cats. *Physiology and Behavior*, **20**, 523–31.

Zagrodzka, J. & Fonberg, E. (1997). Is predatory behavior a model of complex forms of human aggression? In *Aggression: biological, developmental, and social perpectives*, ed. S. Feshbach & J. Zagordzka, pp. 15–27. New York: Plenum Press.

Zahavi, A. (1975). Mate-selection – a selection for a handicap. *Journal of Theoretical Biology*, **53**, 205–14.

Zahavi, A. (1977). The cost of honesty (further remarks on the handicap principle). *Journal of Theoretical Biology*, **67**, 603–5.

Zahavi, A. (1995). Altriuism as a handicap – the limitations of kin selection and reciprocity. *Journal of Avian Biology*, **26**, 1–3.

Zahavi, A. & Zahavi, A. (1997). *The handicap principle: a missing piece of Darwin's puzzle*. New York: Oxford University Press.

Zajonc, R. (1968). Attitudinal effects of mere exposure. *Journal of Personality and Social Psychology*, **9**, 1–27.

Zarrow, M. X., Denenberg, V. H. & Anderson, C. O. (1965). Rabbit: frequency of suckling in the pup. *Science*, **150**, 1835–6.

Zenone, P. G., Sims, M. E. & Erickson, C. J. (1979). Male ring dove behaviour and the defence of genetic paternity. *American Naturalist*, **114**, 615–26.

Zeyl, C. & Bell, G. (1997). The advantage of sex in evolving yeast populations. *Nature*, **388**, 465–8.

Zhang, D. M., Bula, W. & Stellar, E. (1986). Brain cholecystokinin as a satiety peptide. *Physiology and Behavior*, **36**, 1183–6.

Zigmond, M. J., Bloom, F. E., Landis, S. C., Roberts, J. L. & Squire, L. R. (eds.) (1999). *Fundamental neuroscience*. San Diego: Academic Press.

Zimmerberg, B. & Mickus, L. A. (1990). Sex differences in corpus callosum: influence of prenatal alcohol exposure and maternal undernutrition. *Brain Research*, **537**, 115–22.

Zuckerman, M. (1984). Sensation seeking: a comparative approach to a human

trait. *Behavioral and Brain Sciences,* **7**, 413–71.
Zuckerman, M. (1994). *Behavioral expressions and biosocial bases of sensation seeking.* New York: Cambridge University Press.
Zuk, M., Thornhill, R., Ligon, J. D. & Johnson, K. (1990). Parasites and mate choice in red jungle fowl. *American Zoologist,* **30**, 235–44.

Author index

Subject index